COMPUTER VISION AND RECOGNITION SYSTEMS

Research Innovations and Trends

COMPUTER VISION AND RECOGNITION SYSTEMS

Research Innovations and Trends

Edited by
Chiranji Lal Chowdhary, PhD
G. Thippa Reddy, PhD
B. D. Parameshachari, PhD

First edition published 2022

Apple Academic Press Inc.
1265 Goldenrod Circle, NE,
Palm Bay, FL 32905 USA

4164 Lakeshore Road, Burlington,
ON, L7L 1A4 Canada

CRC Press
6000 Broken Sound Parkway NW,
Suite 300, Boca Raton, FL 33487-2742 USA

2 Park Square, Milton Park,
Abingdon, Oxon, OX14 4RN UK

© 2022 by Apple Academic Press, Inc.

Apple Academic Press exclusively co-publishes with CRC Press, an imprint of Taylor & Francis Group, LLC

Reasonable efforts have been made to publish reliable data and information, but the authors, editors, and publisher cannot assume responsibility for the validity of all materials or the consequences of their use. The authors, editors, and publishers have attempted to trace the copyright holders of all material reproduced in this publication and apologize to copyright holders if permission to publish in this form has not been obtained. If any copyright material has not been acknowledged, please write and let us know so we may rectify in any future reprint.

Except as permitted under U.S. Copyright Law, no part of this book may be reprinted, reproduced, transmitted, or utilized in any form by any electronic, mechanical, or other means, now known or hereafter invented, including photocopying, microfilming, and recording, or in any information storage or retrieval system, without written permission from the publishers.

For permission to photocopy or use material electronically from this work, access www.copyright.com or contact the Copyright Clearance Center, Inc. (CCC), 222 Rosewood Drive, Danvers, MA 01923, 978-750-8400. For works that are not available on CCC please contact mpkbookspermissions@tandf.co.uk

Trademark notice: Product or corporate names may be trademarks or registered trademarks and are used only for identification and explanation without intent to infringe.

Library and Archives Canada Cataloguing in Publication

Title: Computer vision and recognition systems : research innovations and trends / edited by Chiranji Lal Chowdhary, PhD, G. Thippa Reddy, PhD, B. D. Parameshachari, PhD.
Names: Chowdhary, Chiranji Lal, 1975- editor. | Reddy, G. Thippa, editor. | Parameshachari, B.D., 1981- editor.
Description: First edition. | Includes bibliographical references and index.
Identifiers: Canadiana (print) 20210317345 | Canadiana (ebook) 20210317388 | ISBN 9781774630150 (hardcover) | ISBN 9781774639368 (softcover) | ISBN 9781003180593 (ebook)
Subjects: LCSH: Computer vision. | LCSH: Artificial intelligence. | LCSH: Pattern recognition systems.
Classification: LCC TA1634 .C66 2022 | DDC 006.3/7—dc23

Library of Congress Cataloging-in-Publication Data

CIP data on file with US Library of Congress

ISBN: 978-1-77463-015-0 (hbk)
ISBN: 978-1-77463-936-8 (pbk)
ISBN: 978-1-003180-59-3 (ebk)

About the Editors

Chiranji Lal Chowdhary, PhD, is Associate Professor in the School of Information Technology & Engineering at VIT University, India, where he has been since 2010. From 2006 to 2010 he worked at the M.S. Ramaiah Institute of Technology in Bangalore, India, eventually as a lecturer. His research interests span both computer vision and image processing. Much of his work has been on images, mainly through the application of image processing, computer vision, pattern recognition, machine learning, biometric systems, deep learning, soft computing, and computational intelligence. As of 2020, Google Scholar reports over 400+ citations to his work. He has given several invited talks on medical image processing. Professor Chowdhary is editor/co-editor of three books and is the author of over 40 articles on computer science. He filed two patents deriving from his research.

Dr. Chowdhary received a BE (CSE) from MBM Engineering College at Jodhpur, India, in 2001, and MTech (CSE) from the M. S. Ramaiah Institute of Technology at Bangalore, India, in 2008. He received his PhD in Information Technology and Engineering from the VIT University Vellore in 2017.

Google Scholar: https://scholar.google.com/citations?user=PpJt13oAAAAJ&hl=en
ORCID ID: https://orcid.org/0000-0002-5476-1468

G. Thippa Reddy, PhD, is currently working as Associate Professor in School of Information Technology and Engineering, VIT, Vellore, Tamil Nadu, India. He has more than 14 years of experience in teaching. He has published more than 50 international/national publications. Currently, his areas of research include machine learning, Internet of Things, deep neural networks, blockchain, and computer vision. He has filed one patent deriving from his research.

He obtained his BTech in CSE from Nagarjuna University, India; MTech in CSE from Anna University, Chennai, Tamil Nadu, India; and his PhD at VIT, Vellore, Tamil Nadu, India.

Google Scholar: https://scholar.google.com/citations?user=nQFCxmkAAAAJ&hl=en&oi=ao
Researchgate: https://www.researchgate.net/profile/Thippa_Gadekallu

B. D. Parameshachari, PhD, is Professor and Head in the Department of Telecommunication Engineering at GSSS Institute of Engineering & Technology for Women, Mysuru Affiliated to Visvesvaraya Technological University (VTU), Belagavi, Karnataka, India. Under his leadership, the Deptartment of Telecommunication Engineering at GSSSIETW has achieved NBA (Tier-II) accreditation twice. He was instrumental in establishing collaboration between GSSSIETW and Multimedia University, Malaysia, and also with University of Sannio, Italy.

Dr. Parameshachari has over 17 years of teaching and research experience, and he has worked in various positions and places, including Karnataka, Kerala, and Mauritius (abroad). He is recognized as a Research Guide at VTU, Belagavi, and currently five research scholars are pursuing PhD degrees under his supervision.

Dr. Parameshachari has completed his BE degree in Electronics and Communication Engineering at Kalpatharu Institute of Technology, Tiptur, India, his MTech degree in Digital Communication Engineering at BMS College of Engineering, Bangalore, and his PhD in ECE from Jain University, Bangalore, India.

More details can be found on his webpage:
http://www.geethashishu.in/te/item/184-parameshachari-b-d

Contents

Contributors .... *ix*
Abbreviations .... *xi*
Preface .... *xv*

1. **Visual Quality Improvement Using Single Image Defogging Technique**... 1
   Pritam Verma and Vijay Kumar

2. **A Comparative Study of Machine Learning Algorithms in Parkinson's Disease Diagnosis: A Review** 13
   Pedram Khatamino and Zeynep Orman

3. **Machine Learning Algorithms for Hypertensive Retinopathy Detection through Retinal Fundus Images** 39
   N. Jagan Mohan, R. Murugan, and Tripti Goel

4. **Big Image Data Processing: Methods, Technologies, and Implementation Issues** 69
   U. S. N. Raju, Suresh Kumar Kanaparthi, Mahesh Kumar Morampudi, Sweta Panigrahi, and Debanjan Pathak

5. **N-grams for Image Classification and Retrieval** 93
   Pradnya S. Kulkarni

6. **A Survey on Evolutionary Algorithms for Medical Brain Images** 121
   Nurşah Dincer and Zeynep Orman

7. **Chatbot Application with Scene Graph in Thai Language** 149
   Chantana Chantrapornchai and Panida Khuphira

8. **Credit Score Improvisation through Automating the Extraction of Sentiment from Reviews** 165
   Aadit Vikas Malikayil, Maheswari R., Azath H., and Sharmila P.

9. **Vision-Based Lane and Vehicle Detection: A First Step Toward Autonomous Unmanned Vehicle** 183
   Tapan Kumar Das

10. **Damaged Vehicle Parts Recognition Using Capsule Neural Network** 197
    Kundjanasith Thonglek, Norawit Urailertprasert, Patchara Pattiyathanee, and Chantana Chantrapornchai

11. **Partial Image Encryption of Medical Images Based on Various Permutation Techniques** 223
    Kiran, B. D. Parameshachari, H. T. Panduranga, and Rocío Pérez de Prado

12. **Image Synthesis with Generative Adversarial Networks (GAN)** 239
    Parvathi R. and Pattabiraman V.

Index *251*

Contributors

H. Azath
VIT Bhopal, India

Chantana Chantrapornchai
Faculty of Engineering, Kasetsart University, Bangkok, Thailand

Chiranji Lal Chowdhary
School of Information Technology & Engineering, VIT Vellore, Tamil Nadu, India

Tapan Kumar Das
School of Information Technology and Engineering, Vellore Institute of Technology, Vellore, India

Rocío Pérez de Prado
Linares School of Engineering, Telecommunication Engineering Department, Scientific-Technical Campus of Linares-University Ave. Linares (Jaén), Spain

Nurşah Dincer
Department of Computer Programming, School of Advanced Vocational Studies, Dogus University, 34680 Istanbul, Turkey

Tripti Goel
Department of Electronics and Communication Engineering, National Institute of Technology Silchar, Assam 788010, India

Kiran
Department of ECE Engineering, Vidyavardhaka Engineering College, Mysuru, India

Suresh Kumar Kanaparthi
Department of Computer Science and Engineering, National Institute of Technology Warangal, Telangana State, India

Pedram Khatamino
Department of Computer Engineering, İstanbul University - Cerrahpaşa, İstanbul, Turkey

Panida Khuphira
Faculty of Engineering, Kasetsart University, Bangkok, Thailand

Pradnya S. Kulkarni
School of Computer Engineering and Technology, MIT World Peace University, Pune, India
Honorary Research Fellow, Federation University, Australia

Vijay Kumar
Department of Computer Science and Engineering, National Institute of Technology, Hamirpur, Himachal Pradesh, India

R. Maheswari
VIT Chennai, India

Aadit Vikas Malikayil
VIT Chennai, India

N. Jagan Mohan
Department of Electronics and Communication Engineering, National Institute of Technology Silchar, Assam 788010, India

Mahesh Kumar Morampudi
Department of Computer Science and Engineering, National Institute of Technology Warangal, Telangana State, India

R. Murugan
Department of Electronics and Communication Engineering, National Institute of Technology Silchar, Assam 788010, India

Zeynep Orman
Department of Computer Engineering, İstanbul University-Cerrahpaşa, İstanbul, Turkey

H. T. Panduranga
Department of ECE Engineering, Govt. Polytechnic, Turvekere, Tumkur, India

Sweta Panigrahi
Department of Computer Science and Engineering, National Institute of Technology Warangal, Telangana State, India

B. D. Parameshachari
GSSS Institute of Engineering & Technology for Women, Mysuru, India

R. Parvathi
School of Computer Science and Engineering, Vellore Institute of Technology, Chennai, India

Debanjan Pathak
Department of Computer Science and Engineering, National Institute of Technology Warangal, Telangana State, India

V. Pattabiraman
School of Computer Science and Engineering, Vellore Institute of Technology, Chennai, India

Patchara Pattiyathanee
Kasetsart University, Bangkok, Thailand

U. S. N. Raju
Department of Computer Science and Engineering, National Institute of Technology Warangal, Telangana State, India

G. Thippa Reddy
School of Information Technology & Engineering, VIT Vellore 632014, Tamil Nadu, India

P. Sharmila
Sri Sai Ram Engineering College, India

Kundjanasith Thonglek
Nara Institute of Science and Technology, Nara, Japan

Norawit Urailertprasert
Vidyasirimedhi Institute of Science and Technology, Rayong, Thailand

Pritam Verma
Department of Computer Science and Engineering, National Institute of Technology, Hamirpur, Himachal Pradesh, India

Abbreviations

2DPCA	two-dimensional PCA
AANN	adaptive artificial neural network
ABC	artificial bee colony
ACDE	automatic clustering
ACRO	adaptive coral reef optimization
AHE	adaptive histogram equalization
AIaaS	artificial intelligence as a service
ANN	artificial neural networks
ARCBBO	adaptive real-coded biogeography-based optimization
ASNR	average signal to noise ratio
ASSO	adaptive swallow swarm optimization
AWDO	applied adaptive wind-driven optimization
BBO	biogeography-based optimization
BCP	bright channel prior
BDB	Bayesian detection boundaries
BF	bilateral filter
BFA	bacterial foraging algorithm
BFO	bacterial foraging optimization
BID	big image data
BIDP	big image data processing
BoVW	bag-of-visual-words
BP	blood pressure
BPSO	binary particle swarm optimization
BP-NN	back propagation neural network
BV	blood vessels
CBIR	content-based image retrieval
CBOW	contextual-bag-of-words
CII	contrast improvement index
CLAHE	contrast limited adaptive histogram equalization
CNN	convolutional neural networks
CPP	curve partitioning points
CRAE	central retinal arterial equivalent
CSF	cerebrospinal fluid

CSA	crow search algorithm
CSA	cuckoo search algorithm
CWS	cotton wool spots
DBCP	double bright channel prior
DCP	dark channel prior
DCT	discrete cosine transform
DE	differential evolution
DICP	dynamic ICP
DNN	deep neural networks
DP	dynamic pruning
DPSO	Darwinian particle swarm optimization
DR	diabetic retinopathy
DR2T	discrete ripplet-II transforms
DSA	differential search algorithm
DWI	diffusion-weighted imaging
DWT	discrete wavelet transform
EA	evolutionary algorithms
EP	evolutionary programming
ELM	extreme learning machine
ES	evolution strategy
FA	firefly algorithm
FCM	fuzzy c-means
FLAIR	fluid-attenuated inversion recovery
FODPSO	fractional-order DPSO
FOV	field of view
FSVM	fuzzy SVM
GA	genetic algorithm
GAN	generative adversarial network
GEP	gene expressing programming
GET	generic edge tokens
GM	gray matter
GP	genetic programming
GRNN	general regression neural network
GWO	gray wolf optimizer
HGOA	histogram-based gravitational optimization algorithm
HOG	histogram of oriented edges
HR	hypertensive retinopathy
HVS	human visual system

Abbreviations

ICP	image cloud processing
KC	Kappa coefficient
KFECSB	kernelized fuzzy entropy clustering with local spatial information and bias correction
KNN	K-nearest neighbor
KSVM	kernel SVM
LBP	local binary pattern
LDA	linear discriminant analysis
LTP	local ternary pattern
MA	micro aneurysms
MCET	minimum cross entropy thresholding
MCMAR	multiagent-consensus-MapReduce-based attribute reduction
MDCS	Matlab Distributed Computing Server
ME	measure of enhancement
MFCM	modified fuzzy c-means
MFKM	modified fuzzy k-means algorithm
ML	machine learning
MPSO	modified particle swarm optimization
MRT	mean relative tremor
MSE	mean square error
NB	Naïve Bayes
NCA	number of changes in acceleration
NCV	number of changes in velocity direction
NIFCMGA	neighborhood intuitionistic fuzzy c-means clustering algorithm with a genetic algorithm
OPF	optimum path forest
PBDS	pathological brain detection system
PCA	principal component analysis
PD	Parkinson's disease
PNN	probabilistic neural network
PSNR	peak signal to noise ratio
PSO	particle swarm optimization
QPSO	quantum-behaved particle swarm optimization
RBF	radial basis function
RDD	resilient distributed dataset
RNN	recurrent neural network
RST	rough set theory

SBS	sequential backward selection
SFS	sequential forward selection
SICP	static ICP
SIFT	scale-invariant feature transform
SNR	signal to noise ratio
SOM	self-organizing map
SR	stochastic resonance
SSIM	structural similarity index metric
STRSPSO-RR	supervised tolerance rough set–PSO-based relative reduct
STRSPSO-QR	supervised tolerance rough set–PSO-based quick reduct
SVM	support vector machine
SURF	speeded up robust features
TGIA	Thai General Insurance Association
TRS	tolerance rough set
TRSFFQR	tolerance rough set firefly-based quick reduct
UCI	University of California-Irvine
VAE	variation auto-encoder
WM	white matter
WOA	whale optimization algorithm

Preface

This volume, *Computer Vision and Recognition Systems: Research Innovations and Trends,* is the contribution of authors from Thailand, Spain, Japan, Turkey, Australia, and India. The focus of the volume is based on essential modules for comprehending all artificial intelligence experiences to provide machines with the power of vision. To imitate human sight, the computer vision needs to obtain, store, interpret, and understand images.

Despite its incredible growth in neural networks, machine learning, and deep learning, surprisingly very few books are available on these aspects of the topics in the form of research contributions of computer vision and recognition systems. The main objectives of this book are to provide innovative research developments, applications, and current trends in computer vision and recognition systems.

We are thankful to our contributors for quality submissions based on various research works such as visual quality improvement, Parkinson's disease diagnosis, hypertensive retinopathy detection through retinal fundus, big image data processing, N-grams for image classification, medical brain images, chatbot application, credit score improvisation, vision-based lane vehicle detection, damaged vehicle parts recognition, partial image encryption of medical images, and image synthesis.

The content is presented in chapter format, and the organization of these chapters formed this book to provide support in various areas where computer vision is being used. Computer vision helps computers to perceive the images and to label them. The subject area include different approaches to computer vision, image processing, and frameworks for machine learning to build automated and stable applications. Deep learning is also included for making immersive application-based chapters, pattern recognition, and biometric systems.

CHAPTER 1

Visual Quality Improvement Using Single Image Defogging Technique

PRITAM VERMA and VIJAY KUMAR*

Department of Computer Science and Engg., National Institute of Technology, Hamirpur, Himachal Pradesh, India

Corresponding author. E-mail: vijaykumarchahar@gmail.com

ABSTRACT

In the wintry weather period, haze is the prime confront during driving. It eliminates the visibility of an image. Fog removal techniques are required to improve the visibility level of the image. In this chapter, a hybrid approach is implemented for fog removal. The expected approach utilizes the basic concepts of Dark Channel Prior and Bright Channel Prior. Apart from this, order statistic filter would use to refine the transmission map. The bright channel prior to boundary constraints would use to restore the edges. The proposed technique has been compared with existing techniques over a set of well-known foggy images. The proposed approach outperforms the predefined techniques in terms of average gradient and percentage of saturated pixels.

1.1 INTRODUCTION

Additional climate-related incidents are happened due to fog. In the year 2016, around 9000 peoples were died due to intense fog. The visual quality of images[2] is ruined due to the being there of dust, smoke, etc. Differences between fog, haze, and rain are described in Table 1.1.[15] The core cause for fog in the environment due to water droplets suspension.[1,3] The water droplets are the reason for consumption and dispersion. When the light comes toward the camera or the viewer is incapacitated due to scattering through droplets and distort the visual quality of the image.[4,6,7] To conquer this problem, some sophisticated systems have been developing to maximize visibility during restraining the strong and dazzling light for oncoming vehicles.[12] For the recognition of fog the motor vehicle detection system was developed[8,13,21] but the main tribulations would have occurred that could not be able to remove the sky visibility. The automatic fog detection could detect only daytime fog but it would not able to detect the nighttime fog. To conquer this problem, computer vision techniques have been started to use.[11,14] These techniques also helped to cut down the operating cost and accommodated a better visual system.[10,25] He et al.[16] planned a Dark Channel Prior (DCP) that would have utilized image pixels with low-intensity value in at least one of the color channels. Nevertheless, this value could be lessened in contrast due to additive air light. DCP commonly use to evaluate the transmission map and atmosphere shroud.[9,20]

TABLE 1.1 Weather Conditions and the Corresponding Particle Size.

Condition	Type	Radius (in μm)	Concentration (cm^{-3})
Fog	Water droplet	1–10	100–10
Haze	Aerosol	10^{-2}–1	10^3–10
Rain	Water droplet	10^2–10^4	10^{-2}–10^{-5}
Cloud	Water droplet	1–10	300–10

Fattal[18] described the local color line prior to re-establish hazy images. Nandal and Kumar (2018) proposed a novel image defogged model that would use fractional-order anisotropic diffusion. They would have used the air light map that would have been evaluated from the hazy model as the picture in the anisotropic dissemination development. However, it went through halo artifacts. To reduce this problem,[19] implemented a technology that would use improved DCP and contrast adaptive histogram equalization that would able to remove the halo artifact with a new median operator in the DCP. They would use a guided filter for the alteration of the transmission map. Contrast Limited Adaptive Histogram Equalization (CLAHE) would use for further visibility improvement but the complexity of computational was so high. To cut down the complexity problem,[22] integrated DCP and Bright channel prior (BCP) would have been developed. They would use BCP to solve the sky-region problem that would relate with DCP-based dehazing.[5] They would use gain intervention filter to increase the computation speed and improve edge preservation. In spite of this, this technique would not able to provide the optimum solution for degraded images. To reduce the above-mentioned problem, the hybrid algorithm is implemented that integrates the DCP and BCP. The proposed approach uses a 2D order statistic filter to illuminate the transmission map. BCP with boundary constraints is being used to restore the edges. This technique is being compared with the existing techniques over a set of well-known foggy images. The leftover configuration of this section is as follows. Section 1.2 briefly describes the degradation model. The proposed defogging techniques are mentioned in Section 1.3. Experimental fallout and planning are given in Section 1.4. The concluding observations are given in Section 1.5.

1.2 DEGRADATION MODELS

Mathematical model of a fog image is represented as:

$$obI(x) = hfI(x)e^{(-\partial d[x])} + Air(1 - e^{-\partial d[x]}) \qquad (1.1)$$

here, image coordinates are denoted by x. Observed hazy image is denoted by obI, Haze free image is represented by hfI, Air is the global light, the scattering coefficient is denoted by ∂, and sense depth is d. The transmission is represented as[20]:

$$tra(x) = e^{-\partial d(x)} \qquad (1.2)$$

In the clear weather condition, $\partial \approx 0$. However, ∂ becomes non-negligible for foggy images. First term from the eq.1.1, obI(x)hfI(x) decreases when the depth scene increases and second term, Air(1-tra[x]) increases when the depth scene increases. The main aim of fog removal from an image to recover the hfI from obI. Air and tra can be estimated from obI. hfI can be obtained as[16]:

$$hfI(x) = \frac{obI(x) - Air}{tra(x)} + A \qquad (1.3)$$

1.2.1 DARK CHANNEL PRIOR

According to the DCP, an RGB image has at least one color channel that have some pixels of lowest intensities that tends to zero. For examples, an image of mountains, stones, tree, some brighter objects, etc. In case of some images of mountains, stones will have lowest intensities as compared to brighter objects and sky region of the sky. The dark channel suggests that an RGB image have at least one color channel which has lowest intensities that are almost tends to zero. Dark Channel mathematically represented as[16]:

$$hfI^{dark}(x) = \min_{m,n \in \lambda(m,n)} \min_{c \in (R,G,B)} hfI^c(y) \quad (1.4)$$

where hfI^c denotes the intensity of the color channel $c \in (R,G,B)$ of the RGB image and and $\lambda(x)$ is a local patch centered at pixel. The minimum value among the three-color channels and all pixels are considered as the dark channel hfI^{dark}. The dark channel pixel value can be approximated as follow[16]:

$$hfI^{dark} \approx 0 \quad (1.5)$$

The dark channel is known as DCP when the approximation is zero for the pixel values. Another part of this, the dark channel for the foggy images produces the pixels that have values greater than zero. Global atmosphere light heads to be achromatic and bright. A combination of air light and direct depletion significant increases the minimum value of the three colors in the local patch. This signifies that the pixel values of the dark channel can play a particular rule to estimate the fog density.

1.2.2 DCP-BASED IMAGE DEFOGGING

In DCP-based Image Defogging algorithm, the dark channel formulated from the input image (see eq 1.4). The atmosphere and transmission map is achieved from the dark channel. The transmission map is further refined and fog free image is reformulated using eq 1.3. The degradation mathematically represented as[6]:

$$obI(x) = hfI(x)e^{(-\partial d(x))} + Air\left(1 - e^{-\partial d(x)}\right) \quad (1.6)$$

To get minimum intensity in the local patch of each color is done by dividing both side of eq. 1.6 by Air^c as follow:

$$\min\left(\frac{obI^c(x)}{Air^c}\right) = \overline{tra(x)} \min \frac{}{Air^c} + \left(1 - \overline{tra(x)}\right) \quad (1.7)$$

Then, the min operator of the three color channel is applied to eq. 1.7 as follow:

$$\min\left(\min\left(\frac{obI^c(x)}{Air^c}\right)\right) = \overline{tra(x)} \min\left(\min\left(\frac{hfI^c}{Air^c}\right)\right) + (1 - \overline{tra(x)}) \quad (1.8)$$

$\overline{tra(x)}$ can be evaluated as

$$\overline{Tr}^{dark}(x) = 1 - \min\left(\min\left(\frac{obI(x)}{Air^c}\right)\right) \quad (1.9)$$

The dark channel pixels value is highly associated with fog density. Therefore, the 0.1% of the brightest pixels in the dark will be selected and the color with highest intensity value among the choose pixels have been used as the value for Air. For the sky region DCP is not reliable. If the color of the sky is close to Air in hazy image then, min(min(obI(x)Air^c)) will approx. to 1 and tra(x)^c will be 0. Haze free image can be mathematically represented for given Air, tra(x)^c and obI(x), as[18]:

$$hfI(x) = \left(\frac{obI(x) - Air}{max(\overline{tra(x)}, p_0)}\right) + Air \quad (1.10)$$

here, the lower bound for transmission is denoted by p_0.

1.2.3 BRIGHT CHANNEL PRIOR

In case of stumpy illumination color images, image enrichment technology is frequently used. A large amount of the brightness augmentation algorithm depended on the BCP that center of attention is the gray removal. The local patches in elucidation images are full of some pixels that have very high intensities in at least one color channel. The construction of a fog picture is definite as:

$$obI(x) = hfI(x)e^{(-\partial d(x))} + Air(1 - e^{-\partial d(x)}) \quad (1.11)$$

where x represents the image coordinates, obI is the observed hazy image, hfI is the haze-free image, Air is the global atmospheric light, ∂ is the scattering coefficient of the atmosphere, and d is the scene depth. The transmission map is defined as:

$$tra(x) = e^{-\partial d(x)} \quad (1.12)$$

The deformation is given as:

$$\left(\frac{obI(x)}{Air^c}\right) = tra(x)\left(\frac{hfIc}{Air^c}\right) + (1 - tra(x)) \quad (1.13)$$

where $c \in \{r,g,b\}$ is the color channel index. We calculate the bright channel on both sides of eq. 1.13. The maximum operators are applied on both sides.

$$\max_{y \in \lambda}\left(\max \frac{obI(y)}{Air^c}\right) = \overline{tra(x)} \max_{y \in \lambda}\left(\max \frac{hfIc}{Air^c}\right) + 1 - \overline{tra(x)} \quad (1.14)$$

where hfI^c is a color channel of hfI and $\lambda(x)$ is a local patch centered at x. We assume that the patch's transmission is $tra(x)$. The goal of this model is to recover hfI, Air, and tra from obI. The low illumination images is defined as[17]:

$$hfI^{bright}(x) = \max_{m,n \in \lambda(m,n)} \left(\max_{c \in (R,G,B)} hfI^c(y)\right) \quad (1.15)$$

$$\overline{Tr}^{bright}(x) = 1 - hfI^{bright}(x) \quad (1.16)$$

here, $hfI^{bright}(x)$ is the bright channel.

1.3 PROPOSED DEFOGGING ALGORITHM

The proposed defogging algorithm is inspired from the work done by Singh and Kumar. The improvements in the work proposed by Singh and Kumar are as follows:

a) Double BCP (DBCP) is used instead of single BCP. The reason behind is to solve the sky-region problem. One BCP is computed by utilizing the boundary constraints and another BCP is computed by utilizing the pad image.
b) 2D order statistic filter is used to preserve the edge information and allow defogging in smooth area.
c) DCP with boundary constraints is used. The atmosphere light and integrated transmission map is estimated by using DCP with boundary constraints and DBCP.

1.3.1 BOUNDARY CONSTRAINTS

It is a lower and upper bound limit of the solution *x*. By the help of this, faster and reliable solutions can be generated by holding the upper and lower bounds limit. Let's consider that bounds are vector with the same length as *x*.

- If no lower bound for any component then use -Inf as the bound and use Inf for no upper bound.
- If either have upper or lower bound, then don't need to write the other type. For example, if have no upper bounds then do not need to supply the other vector of Infs.
- Out of n component, if the first m have bounds then they have to supply a vector of length m containing bounds.

For example, their boundaries are $x \geq 7$ and ≤ 3. The constraint vectors can be *lb* = lower-bound= [-Inf;-Inf;7] and upper-bound = [Inf;3] (will give a warning) or upper bound = [Inf;3;Inf].

1.3.2 DCP WITH BOUNDARY CONSTRAINTS

This is used to eliminate the fog from the foggy image. DCP uses patch wise transmission form boundary constraints. It uses the hazy image, air light, and check pixel-wise boundary for each colour (RGB) and uses the max filter on concentration for the result set of the RGB.

1.3.3 BCP WITH BOUNDARY CONSTRAINTS

Basically, it is used to eliminate the sky-region problem from the foggy image. BCP uses patch-wise transmission form boundary constraints. It uses the hazy image, air light, and check pixel-wise boundary for each color (RGB) and uses the max filter on concentration for the result set of the RGB.

1.3.4 BCP WITH PAD IMAGE

Pad-size (array A) is a vector of no-negative integers that determines both the padding amount and the dimension along which is to add it. The amount of padding is determined the value of an element in the vector. The dimension along which to add the padding is determined the order of an element in vector specifies. By using pad array, it finds the pad-size and pad image. BCP uses the both hazy image and frame size. Maximum patch or brighter pixel from the image is given by this. For the better visual quality of a recovered image, this algorithm is using BCP with boundary constraints.

1.3.5 ALGORITHM

1. Take foggy image as an input.
2. Estimate Air light using 2D order statistical filter.
3. Apply DCP using boundary constraints from eq. 1.4.
4. Estimate the transmission map that is given by eq. 1.9.
5. Apply BCP with boundary constraints from eq. 1.16.
6. Apply BCP with pad image and estimate transmission map.
7. Integrated the both transmission maps obtained from BCP and DCP.

$$\overline{Tr}^{integrated}(x) = \left(\overline{Tr}^{dark}(x)\right) / \left(\overline{Tr}^{bright}(x)\right) \quad (1.17)$$

8. Passed-integrated transmission map into defogged model:

$$hfI_{final}(x) = Air^{ordfilt2}(x) + \left(\frac{hfI - Air^{ordfilt2}(x)}{\max(\overline{Tr}^{integrated}(x), p_0)}\right) \quad (1.18)$$

here, p_0 represents the lower bound.

1.4 EXPERIMENTAL RESULTS

In order to certify the performance of proposed technique, it would compare with the existing dehazing algorithms over 20 images.

1.4.1 EXPERIMENTAL SETUP

This section presents the assessment of the proposed method on MATLAB 9.0, 64-bitIntelR©CoreTMi3-5005U processor with memory of 4 GB. To compare the performance of the proposed defogging technique, benchmark foggy images namely Canon, Toys, Pumpkins, and Cones are taken from well-known SPOT database. The pros defogging technique is compared with four well-known techniques namely.[16,22–24]

1.4.2 QUANTITATIVE ANALYSIS

The performance of the proposed Hybrid technology is evaluated in terms of saturated and average gradient. The value of saturated pixels should be minimum and average gradient should be maximum for the better visual quality. The values of these performance metrics are measured in terms of "mean standard deviation." Table 1.2 show the results obtained from the proposed dehazing technique and other compared the algorithm in terms of saturated. It is observed from the Table 1.2 that illustrates the results required from the implemented defogging technology has less number of saturated pixels than the competitive algorithms. Table 1.3 demonstrates the average gradient occurred from the implemented technique and other compared algorithm. It could be seen from Table 1.3 that is proposed technique preserves the edges as compared to other algorithms.

TABLE 1.2 Performance Comparison in Terms of Saturated Pixels.

Image	[16]	[23]	[24]	[22]	Proposed Approach
Image 1	0.99 ± 0.06	0.97 ± 0.04	0.82 ± 0.05	0.52 ± 0.04	0.47 ± 0.05
Image 2	0.97 ± 0.03	0.81 ± 0.06	0.74 ± 0.04	0.59 ± 0.03	0.52 ± 0.02
Image 3	0.85 ± 0.04	0.72 ± 0.05	0.70 ± 0.07	0.54 ± 0.06	0.49 ± 0.07
Image 4	0.89 ± 0.08	0.85 ± 0.07	0.81 ± 0.09	0.55 ± 0.07	0.50 ± 0.04

TABLE 1.3 Performance Comparison in Terms of Average Gradient.

Image	[16]	[23]	[24]	[22]	Proposed Approach
Image 1	1.20 ± 0.04	1.28 ± 0.07	1.34 ± 0.06	1.57 ± 0.07	1.89 ± 0.02
Image 2	1.34 ± 0.05	1.53 ± 0.09	1.67 ± 0.08	1.72 ± 0.05	1.90 ± 0.03
Image 3	1.15 ± 0.03	1.31 ± 0.06	1.45 ± 0.04	1.51 ± 0.08	1.97 ± 0.04
Image 4	1.32 ± 0.06	1.57 ± 0.04	1.62 ± 0.05	1.69 ± 0.04	1.82 ± 0.03

1.4.3 QUALITATIVE ANALYSIS

The Qualitative analysis of proposed algorithm would does on benchmark foggy images. Figure 1.1 illustrates the process performed by the proposed technique. It is recognized from the Figure 1.1 that proposed technique is able to eliminate the fog and preserve the edges.

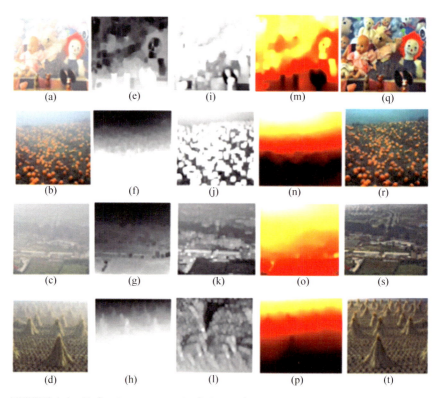

FIGURE 1.1 Defogging process: (a–d) Foggy images, (e–h) dark channel prior, (i–l) double bright channel prior, (m–p) integrated transmission maps, and (q–t) final defogged image.

1.5 CONCLUSIONS

In this chapter, a hybrid defogging technique is proposed that integrated the basic concepts of DCP and BCP. The proposed technique uses DBCP instead of single BCP. The 2D order statistics filter is used to preserve the edge information. The proposed technique is tested on 10 well-known benchmark foggy images. Results reveal that the proposed technique outperforms the existing techniques in terms of ratio of average gradient and saturated pixels by 1.0638 and 1.6931%, respectively. It is able to resolve the sky-region problems that are associated with DCP. The proposed algorithm also removes the halo and artifacts effects from the restored image.

KEYWORDS

- **dark channel prior**
- **bright channel prior**
- **filter**
- **defogging**

REFERENCES

1. Singh, D.; Kumar, V. A Comprehensive Review of Computational Dehazing Techniques. *Ann. Comput. Method Eng*. **2019**, *26* (5), 1395–1413.
2. Singh, D.; Kumar, V.; Kaur, M. Single Image Dehazing Using Gradient Channel Prior. *Appl. Intell.* **2019**, *49* (12), 4276–4293.
3. Chen, B.-H.; Huang, S.-C.; Li, C.-Y.; Kuo, S.-Y. Haze Removal Using Radial Basis Function Networks for Visibility Restitution Application. IEEE Trans. Neural Netw. Learn. Syst. **2018**, *29* (8), 3828–3838.
4. Zhao, H.; Xiao, C.; Yu, J.; Xu, X. Single Image Fog Removal Based on Local Extrema. *IEEE/CAA J. Automatica Sinica* **2015**, *2* (2), 158–165.
5. Singh, D.; Kumar, V. Modified Gain Intervention Filter Based Dehazing Technique. *J. Modern Optics* **2017**, *64* (20), 14–27.
6. Singh, D.; Kumar, V. A Novel Dehazing Model for Remote Sensing Images. *Comput. Electr. Eng.* **2018**, *69*, 14–27.
7. Singh, D.; Kumar, V.; Kaur, M. Image Dehazing Using Window-based Integrated Means Filter. *Multimedia Tools App.* **2019**, 1–23.
8. Parimala, M.; RM, S. P.; Reddy, M. P. K.; Chowdhary, C. L.; Poluru, R. K.; Khan, S. Spatiotemporal-based Sentiment Analysis on Tweets for Risk Assessment of Event Using Deep Learning Approach. *J. Softw.: Prac. Exp.* **2020**.
9. Khare, N.; Devan, P.; Chowdhary, C. L.; Bhattacharya, S.; Singh, G.; Singh, S.; Yoon, B. SMO-DNN: Spider Monkey Optimization and Deep Neural Network Hybrid Classifier Model for Intrusion Detection. *Electronics* **2020**, *9* (4), 692.

10. Das, T. K.; Chowdhary, C. L.; Gao, X. Z. Chest X-Ray Investigation: A Convolutional Neural Network Approach. *J. Biomimetics, Biomater. Biomed. Eng.* **2020**, *45*, 57–70.
11. Reddy, T.; RM, S. P.; Parimala, M.; Chowdhary, C. L.; Hakak, S.; Khan, W. Z. A Deep Neural Networks Based Model for Uninterrupted Marine Environment Monitoring. *Comput. Commun.* **2020**.
12. Anwar, M. I.; Khosla, A. Vision Enhancement through Single Image Fog Removal. *Eng. Sci. Technol.* **2017**, *20* (3), 1075–1083.
13. Hautière, N.; Tarel, J.; Lavenant, J.; Aubert, D. Automatic Fog Detection and Estimation of Visibility Distance through Use of an Onboard Camera. *Mach. Vision App.* **2005**, *17*, 8–20.
14. Bronte, S.; Bergasa, L. M.; Alcantarilla, P. F. Fog Detection System Based on Computer Vision Techniques. In *12th International IEEE Conference on Intelligent Transportation Systems* 2009, pp. 1–6.
15. Narasimhan, S. G.; Nayar, S. K. Vision and the Atmosphere. *Int. J. Comput. Vision* **2002**, *48* (3), 233–254.
16. He, K.; Sun, J.; Tang, X. Single Image Haze Removal Using Dark Channel Prior. *IEEE Trans. Pattern Analy. Mach. Intell.* **2011**, *33* (12), 2341–2353.
17. Kaur, M.; Singh, D.; Kumar, V.; Sun, K. Color Image Dehazing Using Gradient Channel Prior and Guided L0 File Information Sciences. 2020.
18. Fattal, R. Dehazing Using Color-lines. *ACM Trans. Graphics* **2014**, *34* (1), 13.
19. Kapoor, R.; Gupta, R.; Son, L.; Kumar, R.; Jha, S. Fog Removal in Images Using Improved Dark Channel Prior and Contrast Limited Adaptive Histogram. 2019.
20. Nandal, S.; Kumar, S. R. Single Image Fog Removal Algorithm in Spatial Domain Using Fractional Order Anisotropic Diffusion. *Multimedia Tools App.* **2018**, *78*, 10717–10732.
21. RM, S. P.; Maddikunta, P. K. P.; Parimala, M.; Koppu, S.; Reddy, T.; Chowdhary, C. L.; Alazab, M. An Effective Feature Engineering for DNN Using Hybrid PCA-GWO for Intrusion Detection in IoMT Architecture. *Comput. Commun.* **2020**.
22. Singh, D.; Kumar, V. Single Image Haze Removal Using Integrated Dark and Bright Channel Prior. *Modern Phys. Lett. B* **2018**, *32* (4), 1–9.
23. Pang, J.; Au, O. C.; Guo, Z. Improved Single Image Dehazing Using Guided Filter. *Asia Pacific Sign. Info. Process. Assoc.* **2011**.
24. Khandelwal, V.; Mangal, D.; Kumar, N. Elimination of Fog in Single Image Using Dark Channel Prior. *Int. J. Eng. Technol.* **2018**, *5* (2), 1601–1606.
25. Wu, M.; Zhang, C.; Jiao, Z.; Zhang, G. Improvement of Dehazing Algorithm Based on Dark Channel Priori Theory. *Optik* **2020**, *206*, 164174.

CHAPTER 2

A Comparative Study of Machine Learning Algorithms in Parkinson's Disease Diagnosis: A Review

PEDRAM KHATAMINO and ZEYNEP ORMAN*

Department of Computer Engineering, İstanbul University–Cerrahpaşa, İstanbul, Turkey

*Corresponding author. E-mail: ormanz@istanbul.edu.tr

ABSTRACT

This chapter is a comprehensive literature review as a comparative study of machine learning algorithms in Parkinson's disease diagnosis. The recent studies in the literature that are conducted on different datasets containing both handwriting and voice datasets of Parkinson's disease are analyzed. The fact that Parkinson data are mostly suitable for machine learning analysis, this situation triggers the authors' tendency to research this area. The Parkinson detection literature inclines through deep learning algorithms due to the automatic anomaly detection aspect. The recent studies go toward an automated disease detection and classification system. Therefore, this chapter also aims to include papers that are using deep learning methods for Parkinson's disease diagnosis. The authors strongly believe that it will be a handbook for researchers who are eager to accomplish research on this subject and it will be very beneficial.

2.1 INTRODUCTION

Parkinson's disease (PD), a chronic and progressive disease, generally has symptom like shivering, which occurs in most PD patients. Involuntary shaking may happen in the hands, arms, legs, and chin. However, the uncontrolled movement of the thumbs is one of the most common symptoms. Of course, not every handshake is a sign of PD. In order to make this diagnosis, a general check-up is required by experts. The slowing of movements is a quite common symptom of PD.[1] Unfortunately, the patients are unable to perform the necessary daily life movements over time. As they walk, they may see shrinkage in their steps and begin to lean forward. Apart from these, most common symptoms, speech changes, handwriting deterioration, posture deterioration, sudden movements while sleeping and bowel disorders are the other the symptoms of PD.[2]

PD has spread worldwide due to the modern world lifestyle, which is more common in older people. The PD represents the second most common neurodegenerative disorder after Alzheimer's disease.[3] This disease leads to the limitation of the person's speaking skills, tremors in hand movements and movement and muscle problems in general. PD reduces the standard of living of sick people and naturally affects their families. PD is the second most prevalent neurodegenerative disease in the world, affecting approximately 10 million people worldwide.[4] Non-invasive methods are more suitable for these people because most of them are not physically good. The most common non-invasive methods in clinics in the Parkinson area are the handwriting and voice speech tests.[5] The non-invasive techniques are generally referred to as disease diagnosis methods that do not require surgical intervention.

The datasets collected by these non-invasive methods are generally suitable for analyzing by machine learning techniques. There are many studies conducted in the literature on the diagnosis of PD by using different techniques. Since there is no specific rule of machine learning techniques and parameter optimization, the trial and error approaches are often used.[6] Therefore, experiments with different machine learning methods will enrich and improve the literature.

There are many different sorts of articles and researches in Parkinson's literature. Many machine learning methods could be applied to Parkinson datasets. In recent literature, the accuracy parameter is usually used for evaluating the efficiency of the methods. However, there are many different sorts of machine learning performance evaluating methods like f1 score, sensitivity, and confusion matrix.[52–54]

In this chapter, the Background section presents some useful information about voice and handwriting datasets; additionally, this section contains the possible treatments of PD. The Literature Review section includes a summary of the novel and valuable papers with various classification methods. Solutions and Recommendation section introduces the technical analysis of the previous section in order to find the optimum solutions to the classification problems. Finally, the last section states the conclusion of this chapter.

2.2 PARKINSON'S DISEASE

2.2.1 BACKGROUND

In order to understand the literature review section better, this section will focus on the introduction of voice and handwriting datasets that form the basis of the research. Although the methods of collecting datasets are different from each other, they have all attempted their best to represent the differences between people with PD and healthy individuals. In addition, the slowing and delaying treatment methods used in the disease process will be briefly mentioned in this section. Modern medical approaches have made significant progress, especially in the recent period in this area.

2.2.2 PARKINSON'S DISEASE VOICE DATA

As the aforementioned issues, many people with PD will be considerably dependent on the clinical operation. The essential physical visits to the clinic for diagnosing and treatment are stressful for many people with PD. Researches have shown that the most critical symptoms of PD are dysphonia, gait anomaly, and handwriting tremors. By analyzing the literature, it is feasible to claim that approximately 90% of people with PD exhibit some form of vocal deterioration. The voice of people with PD typically has some sound anomalies which are called dysphonia symptoms. The dysphonia symptom is a general term that refers to disorders of voice, and it consists of different aspects such as pathological or functional problems with one's voice.[7]

In the case of PD, the voice will sound husky, tense, or laborious. Sometimes, the patient's voice may become so rustling and abnormal that the listener may have difficulty understanding the patient's speech. However, voice disorders may have been caused by different causes such as vocal nodule-related disorders in vocal cords, or unexpected vocal complications in the post-operation stage, or unexpected ulcers on the vocal cords. Likewise, misuse of voice can lead to vocal disorders, for instance too high or too low usage of voice, or disorders caused by using voice with inadequate breathing support or postural disorders. Some dysphonia seems like a cross between misuse and something physiological.[8]

When the studies analyzing the voice, data were examined, it was determined that the operations performed were generally to detect abnormal characteristics in the voice data signals. In these studies, the speech sounds datasets include standard speech tests which are recorded by a microphone and the data are analyzed by measurement methods (implemented in software algorithms) to detect certain properties of these signals. Table 2.1 illustrates some features of a voice dataset, which becomes a standard in this field.[9] Some preprocessing operations can be performed on these raw data in order to extract discriminative features, or it is possible to prepare the data in the right format for deep learning architecture's automatic feature selection and learning algorithms.

TABLE 2.1 Voice Dataset Features.

Attribute name	Description
MDVP:Jitter(Abs)	Variation in fundamental frequency
Jitter:DDP	Variation in fundamental frequency
MDVP:APQ	Measures of variation in amplitude
Shimmer:DDA	Measures of variation in amplitude
NHR	Ratio of noise to tonal components
HNR	Ratio of noise to tonal components
RPDE	Dynamic complex measurement
DFA	Signal fractal scaling exponent
D2	Dynamic complex measurement
PPE	Non-linear measure of fundamental frequency
Status	The status of the patient (1)—Parkinson's disease, (0)—Healthy

2.2.3 PARKINSON'S DISEASE HANDWRITING DATA

Handwriting tests are one of the most widely used non-invasive methods in recent years. The idea of collecting data from handwriting tests to detect

PD has led this literature to develop in other directions. Additionally, the literature is enriched with image processing and machine learning structures. Meanwhile, the analysis of voice data is based on older and traditional signal processing methods, whereas handwriting tests can be an excellent alternative method for PD diagnosis.[10]

Many researchers use spiral and meander cases to set a standard in the handwriting datasets that are created in more research hospitals around the world. For instance, Figure 2.1 shows several spiral test samples in which (a) and (b) belong to healthy people and (c) and (d) are the drawings taken from people with PD.[11] However, when the literature is examined, it is possible to observe studies consisting of very different structures and methods. Since the collection of handwriting data is more practical, it is more widely used today.

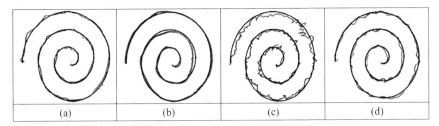

FIGURE 2.1 Spiral test samples.

Handwriting is a complex activity entailing cognitive, kinesthetic, and perceptual-motor components,[12] the changes in which can be a promising biomarker for the evaluation of PD.[13] Indeed, there is evidence to suggest that the automatic discrimination between unhealthy and healthy people can be accomplished based on several features obtained through simple and easy-to-perform handwriting tasks.[14] Developing a handwriting-based decision support system is desirable, as it can provide a complimentary, non-invasive, and very low-cost approach to the standard evaluations carried out by clinical experts.

Using dynamic aspects of the handwriting process helps to create a useful tendency to analyzing potentialities of automatic handwriting systems for PD detection. Several dynamic features of handwriting drawing data are X, Y, Z coordination, pressure, grip angle, and timestamp.[15] For instance, using the features X and Y and their respective pressure features can be useful for the solution of PD classification problem.[16] Dynamic handwriting analysis benefits from the use of digitizing tablets and electronic pens. By using these devices, it is straightforward to measure the temporal and spatial variables of handwriting, the pressure exerted over the writing surface, the pen inclination, and the movement of the pen while not in contact with the surface, etc.

Generally, traditional machine learning, mathematical, statistical and feature selection algorithms such as optimum path forest (OPF), support vector machines (SVM), naive Bayes (NB), gray wolf optimization, cuttlefish optimization, particle swarm optimization, Visual data augmentation, Gaussian mixture model, K-nearest neighbor (KNN), random forest, decision trees, time series-based feature images, artificial neural networks (ANN), self-organizing map (SOM), radial basis function (RBF), linear SVM, Ripper k, fuzzy-KNN, and fuzzy C-means are used for PD handwriting diagnosis. Nowadays, machine learning and deep learning are often used for the classification of medical images that belong to Parkinson patients.[17] However, some different types of ANN or different architectures of convolutional neural networks such as cifar10, ImageNet, LeNet, ResNet, and VGG16 are used for PD handwriting classification.[18]

2.2.4 PARKINSON'S DISEASE TREATMENT

The main goal in the treatment of PD is to enable the patient to become active, independent, and able to do his/her own work. There is no precise treatment for today. However, the limited number of medications used (either provide dopamine, either dopamine-like effect or increase the use of dopamine by inhibiting the disintegration in the brain) is aimed at controlling symptoms. Smart exercise practices, balance exercises, and lifestyle changes can be beneficial. Speech and language therapists may also be helpful in patients with speech disorders. However, if the disease cannot be corrected, the symptoms do not work despite drug use and rehabilitation.

Accordingly, there is no specific medical treatment of PD; the gradual decline of the patient can only be managed during the disease progression. Therefore, it is essential to detect the disease in early stages by machine learning and deep learning methods due to an early diagnosis of PD could be crucial for the prospect of medical treatment; likewise, it is vital for evaluating the effectiveness of new drug treatments at prodromal stages.[19]

2.3 OVERVIEW OF PARKINSON'S DISEASE LITERATURE

2.3.1 LITERATURE REVIEW

Many varieties of machine learning and deep learning methods are deployed for PD detection from voice, gait, and handwriting datasets. For instance, Bernardo et al.[20] introduced a PC app for PD detection. The C# based interface app is designed for capturing data from patients; furthermore, the author developed some algorithms for feature extraction. The author introduced novel samples for a handwritten test like a spiral, triangle, and cube. Several preprocessing algorithms like Color thresholding, RGB convert to grayscale, De-noising the pattern, and Skeletonization process operates for feature extraction. Euclidean distance, relative distance, circular distance, Manhattan distance, mouse pointer speed, the similarity between pixels are features of the dataset. Optimum Path Forest (OPF), SVM and, NB are the classifiers of the research. In this work, the author team reached 100% accuracy with SVM classifier.

Pereira et al.[11] mainly used the preprocessing methods for distinguishing the template and patient drawings from paper-based tests; color thresholding, blur filter, median filter, capturing the pattern of handwritten drawings from the paper-based test are the preprocessing stages of this work. Features like RMS, maximum difference (argmax), minimum difference (argmin), standard deviation, Mean Relative Tremor (MRT) had been extracted from images. From the comparison of OPF, NB, SVM classifiers, SVM classifier reached 67% of accuracy. The authors collected handwriting dataset consists of spirals, meanders, and captured drawings from paper-based tests.

In another research, Pereira et al.[21] designed the extracting method for feature images from handwriting drawings. The author team extended their dataset to six tests such as circle on the paper, circle on the air, diadochokinesis with the right hand, diadochokinesis with the left hand, meander, spiral, and time-series base images. The main purpose of this work was to produce the feature images from raw data by normalizing, squaring, and sketching matrixes into greyscale images as CNN inputs. The data collected by digitized pen from a tablet in which the features were Microphone, Finger grip, Axial Pressure of ink Refill, x, y, z. Different sort of CNN architectures was used such as Cifar10, ImageNet for feature extraction. Classifiers such as OPF, NB, SVM were deployed and had been reached to a 95% accuracy level.

In another work, Harihar et al.[22] designed a hybrid intelligent system for accurate PD diagnosis. Aforementioned work consisted different stages; for instance, feature preprocessing stage was using model-based clustering (Gaussian mixture model), feature reduction/selection stage was using principal component analysis (PCA), linear discriminant analysis (LDA), sequential forward selection (SFS), and sequential backward selection (SBS). In this work, the Parkinson dataset of the University of California-Irvine (UCI) was used, which consisted of voice signals. Voice signal features like MDVP, NHR, HNR, RPDE, D2, DFA, Spread1, Spread2, and PPE were analyzed in this work. Least-square (LS-SVM), probabilistic neural network (PNN), general regression neural network (GRNN) had been used as the classifiers. The full accuracy level (100%) was reached in this work.

Some of the researchers were analyzing just the visual attributes of images in the literature. For instance, Moetesu et al.[23] assessed the visual attributes of handwriting dataset for PD classification. Visual attributes of images had been intensified through the novel approach which could be called a sort of data augmentation; eight task tests were created with using three types of images for the combination of datasets: raw image network, median residual network, and edge image network. The author revealed that the CNN–SVM model reached 83% accuracy in voting decision system. The paper-based dataset contained spiral, l, le, les, lektorka, porovant, nepopadnout, sentence tests.

Drotar's et al.[16] thesis was based on the evaluation of kinematics and pressure Parkinson disease dataset for PD detection. Features of handwriting drawings collected by tablet and smart pen during tests. Tests were composed of drawing an Archimedean spiral, repetitively writing orthographically simple syllables and words, and writing of a sentence. Some useful kinematic features analyzed in this work, for example, stroke speed, speed, velocity, acceleration, jerk, Horizontal velocity/acceleration/jerk, Vertical velocity/acceleration/jerk, Number of changes in velocity direction (NCV), Number of changes in acceleration direction (NCA), Relative (NCV), Relative NCA, On-surface time, Normalization-surface time. KNN, ensemble Adaboost, and SVM classifiers were used for classification, and the highest accuracy percentage was 82% in this work.

For creating novel hybrid models for PD detection, Gupta et al.[24,25] took the advantage of grey wolf and cuttlefish optimization algorithms as the search strategy for feature selection. Modified grey wolf optimization based on updating hunters' positions in an optimum way and optimized cuttlefish algorithm for feature selection consists of four groups: global solution, local search, local solution, random solution were used in this study. Random Forest, KNN, Decision tree classifiers were the applied methods for classification, and 94% accuracy was collected. The main goal of these works was to find the optimal subset of features. Different datasets that were composed of handwriting and voice and gait information of patients were analyzed in this article.

The performance of different architecture of CNN-based models had been evaluated in Pereira et al.[26] The authors designed tests for collecting time-series features of handwriting drawings from patients in order to produce feature images. Features of raw data were composed of Microphone, Finger grip, Axial Pressure of ink Refill, x, y, z, feature images information which were collected by tablet and smartpen. The author used different models of CNN's like cifar10, ImageNet and LeNet and OPF for classification and reached to 85% accuracy. Spiral and meanders were the tasks of the proposed tests.

In another research, Spadoto et al.[27] analyzed the Oxford PD Detection Dataset, which contained voice signals of PD patients. In this work, some preprocessing methods were designed to prepare the dataset for different classifiers like OPF, SVM-RBF, SVM-LINEAR, ANN-MLP, SOM, KNN, and finally reaches 75% accuracy through the analyses. Traditional voice dataset features like MDVP, NHR, HNR, RPDE, D2, DFA, Spread1, Spread2, and PPE were used.

In another article, Diaz et al.[28] proposed the dynamical enhancement of static images of handwriting tests. Dynamically enhanced static image was drawing the points of the samples, instead of linking them; so, by this approach, some kinematic information could be reachable (poral/velocity information). The primary goal of this study was to construct enhanced images from raw data: raw image, median filter, edge images combination for voting classification. The Paper-based dataset composed of some tests such as drawing spiral, l, le, les, lektorka, porovant, nepopadnout, and write a sentence. Different sort of classifiers such as SVM linear, SVM-RFB, RF, ET, ADA were used for the classification and 88% accuracy was obtained.

For instance, in a novel approach, Loconsole et al.[29] modified an EMG signal detector tool for PD detection purposes. The authors collected a handwriting dataset that contained a total of three tasks: sentence and two drawing sample tasks. In this work, some handwriting features are extracted from EMG signals such as Density ratio, Height ratio, Execution time, Execution average Linear speed, Acceleration norm, Gyroscope components, and RMS. Simple ANN model Optimal topology of ANN and SVM was reached to 89% accuracy. Computer vision based handwriting analysis tool and surface ElectroMyoGraphy (sEMG) signal-processing techniques were the central aspects of this work.

In Graça et al.,[30] an online mobile app was designed for data collection of patients. Development of the mobile app for online handwriting tests and also analysis of the gait positions were the main aspects of this work. The mobile app could detect drawings features like Spiral Average Error, Spiral Cross, Spiral Pressure Ratio, Spiral Side Ratio, Tap Time Ratio, and Tap Pressure Ratio. Decision tree, Ripper k, and Bayesian Network classifiers were used to reach 85% of accuracy.

In Drotar et al.,[31] the air movement-based data collection method was used for the Parkinson dataset. Online in-air & on-surface movement-based features were analyzed by SVM classifier, and 85% of accuracy was obtained. There were some spiral and word writing tasks in the applied tests.

Shahbaba et al.[32] proposed a new mathematical approach dpMNL (multinomial logit) for PD classification problems in the voice dataset. The proposed model was using the Dirichlet process mixtures, which allowed maintaining the relationship between the distribution of the response variable and covariates in a non-parametrically way. This model was generative, so it had advantages over the traditional MNL (multinomial logit) models which were discriminative. The five-fold cross-validation method was used for evaluating the performance of the model, and $87.7 \pm 3.3\%$ accuracy was achieved.

In another research, Psorakis et al.[33] investigated the classification ability of the proposed improved mRVMs (multiclass multi-kernel relevance vector machines) over the real world datasets such as Parkinson dataset. The research team achieved some improvements such as convergence measures, sample selection strategies, and model improvements for better results by 10-fold cross-validation with 10 repetitions.

In another work on the Parkinson voice dataset, Little et al.[34] proposed dysphonia detecting for PD detection. Also, the authors proposed a novel dysphonia measure, Pitch Period Entropy (PPE), besides usual speech features. The primary approach of this work was setting the exhaustive search of all possible combinations of dysphonia measures to find the optimum results. The combination of particle swarm optimization algorithm and OPF classifier was also used for classification. As a result of the experiments, the combination of pre-selection filter and exhaustive search with SVM classifier reached 91.4 ± 4.4% of classification accuracy.

Spadoto et al.[35] introduced some evolutionary-based techniques such as Particle swarm optimization, Harmony search algorithm and Gravitational search algorithm for maximizing the OPF classifier performance. The authors analyzed the Oxford PD Detection Dataset through the research. Although OPF classifier reached 71% classification accuracy, the combination of PSO and OPF had been reached 73% accuracy and combinations of HS, GSA, and OPF's results were slightly better than the others (84.01%).

Sakar et al.[36] detected PD from dysphonia measures. The main aspects of this research were selecting the optimum subset of features and building a minimal model bias. Therefore, the authors calculated the relationship between the features and the PD score statically. For this task, the authors utilized maximum-relevance-minimum-redundancy (mRMR) and SVM classifiers. The author used the leave-one-out method for evaluating the generalization level of the proposed model.

Das[37] compared different machine learning approaches as Neural network, DMneural, Regression, and Decision tree for PD classification tasks. Through the experiments, ANN performed better than other models. Different ratios of the dataset were used for the evaluation of the model. As the result of this research, neural networks had achieved 92.9% of classification accuracy when 65% of the dataset was used for training, and 35% of the dataset was used for testing with random data splitting.

Guo et al.[38] suggested a combination of genetic programing and the expectation-maximization algorithm (GP–EM) to transform data of PD dataset. The model was applied to voice dataset with flexible and effective learning modules and Gaussians mixture model of the data. The 10-fold cross-validation was used as a performance validation method of the model. Mean accuracy of GP–EM method was 93.1 ± 2.9%.

Tsanas et al.[39] tried a different combination of feature selection algorithms, and the classifiers were compared to find the best one among them. The feature selection algorithms of this work were the least absolute shrinkage and selection operator (LASSO), minimum redundancy maximum relevance (mRMR), RELIEF, and local learning-based feature selection (LLBFS). Obtained features were classified by random forest and SVM classifiers. By using only 10 dysphonia features, overall accuracy was around 99%.

Astrom et al.[40] introduced a novel approach to use neural networks for medical data processes. The central aspect of this article was to use more than one unique neural network parallelly due to error reduction. The outputs of different sort of NN were evaluated with a rule-based system for weighting the outputs for creating a final decision for PD classification. The designed parallel model allowed the system to learn unlearned data of an NN by another one. In conclusion, the results revealed that the parallel system improved the robustness of the classification procedure. A parallel NN system was composed of nine different NN and it enhanced the ordinary classification rate by almost 10%. The suggested model had achieved 91.20% accuracy.

Chen et al.[41] suggested a fuzzy KNN based system (FKNN) in comparison to SVM. Dataset of this work was a range of biomedical voice measurements obtained from 31 people, and 23 of them with PD. The best classification accuracy (96.07%) obtained by the FKNN based system using a 10-fold cross-validation method could ensure a reliable diagnostic model for the detection of PD. PCA was also used for dimension reduction.

Ozcift[42] analyzed a voice Parkinson disease dataset composed of 31 people, 23 with PD and each person's record has 22 features. The linear SVM was used for selecting the most valuable subset of features (10 features). Through the experiments, three evaluating parameters were considered: accuracy, Kappa Error (KE), and Area under the receiver operating characteristic (ROC). Two base performance measures, IBK (a KNN variant) and KStar (kind of KNN) were used to compare the two main classifiers. By applying RF ensemble to classification, the obtained accuracy was around 97%.

Mandal et al.[43,44] introduced a new dysphonia measure, which was called the severity of the disease. The authors used Haar wavelets as the projection filter, and multinomial logistic regression and linear logistic regression as the classifiers for the research. Feature selection of the study mainly relied on SVM and ranker search methods. The authors compared many conventional approaches in the literature such as Bayesian network, SVM, ANN, Boosting methods, and linear and multinomial logistic regression methods for PD classification. The authors revealed that the study had been reached to 100% of classification accuracy.

Zuo et al.[45] proposed a Particle swarm optimization algorithm for parameter optimization and feature selection. The classifier which was used for this research is a FKNN. The proposed model was a combination of PSO and FKNN; this model's performance had been evaluated through 10-fold cross-validation. The average value of accuracy was 97.47%. A PD voice dataset from UCI database was analyzed through the research.

Luukka et al.[46] introduced a hybrid model for PD detection. The model composed of fuzzy entropy-based feature selection combined with a similarity classifier. This combination proved the efficiency of the model by simplifying the dataset and accelerating the classification process. The model's results revealed that the hybrid model reached a high-level accuracy (mean value = 85.03%).

In another work, Li et al.[47] compared optimization approaches by analyzing different medical datasets. The primary purpose of this work was to find the optimum feature set of datasets for better classification results—a fuzzy-based non-linear transformation method was designed for selecting the optimum feature subset from PD dataset. Also, the authors compared the proposed feature selection method with principal component analysis (PCA) and kernel principal component analysis (KPCA) feature selection methods for illustrating the efficiency of the method. The proposed classification approach was applied on different sorts of datasets such as Pima Indians' diabetes, Wisconsin diagnostic breast cancer, Parkinson disease, echocardiogram, BUPA liver disorders dataset, and bladder cancer cases dataset. In conclusion, fuzzy-based non-linear transformation's performance with SVM classifier was found to be better than other methods (93.47% accuracy).

Ozcift et al.[48] proposed computer-aided diagnosis (CADx) systems to improve accuracy. The author proposed the combination of rotation forest (RF) and some machine learning algorithms (30 ML algorithms) to diagnosis disease from heart, diabetes, and Parkinson's disease datasets. RF classifier predicted the accuracy (ACC), KE, and area under the receiver operating characteristic (ROC) curve (AUC) of 74.47, 80.49, and 87.13% respectively.

Khatamino et al.[49] proposed an efficient convolutional neural network for PD classification. The generalization ability of the model was illustrated by comparing it with conventional machine learning classifiers such as SVM and NB. One of the main purposes of this work was to show the discriminative power of the novel DST test. Another main aspect was to illustrate CNN's flexibility and powerful feature learning ability by comparing it with SVM and NB. However, two main approaches were selected for evaluating the performance of the proposed model (CV,

LOOCV). The proposed model was evaluated well due to it performed effectively on two different handwriting datasets. The model reached to 88.89% classification accuracy in the case of 75% training and 25% testing datasets.

Indira et al.,[9] utilized two methods for PD classification that were the fuzzy C-means clustering and the ANN. Fuzzy C-means clustering method had achieved 68.04% accuracy, 75.34% sensitivity, and 45.83% specificity. Likewise, ANN, which was optimized by the filtering methods and PCA has achieved 92% mean accuracy. PD voice dataset was analyzed in this work.

2.3.2 SOLUTIONS AND RECOMMENDATIONS

This chapter surveyed all studies that used machine learning algorithms in PD diagnosis. In order to analyze the factors affecting the success rate of the proposed algorithms, studies were summarized in terms of the classification methods and classifier types, years, datasets, and accuracy rates as shown in Table 2.2.

As one of the results of this literature review, it is realized that researchers have tended to collect PD data, especially in collaboration with the research hospitals in many studies. It is evident that the ideas and guidance of the doctors and medical experts of the neurology departments are essential.

In general, it is observed that high accuracy percentages have been achieved in the literature in the last few years. One of the significant reasons for this performance growth is the improvements in machine learning and deep learning libraries of different programming languages.

The analysis of handwriting data often shifts to the field of image processing. Therefore, some researchers are trying new methods of deep learning rather than old image processing techniques in current studies. Some studies have shown successful results of the CNN structure. CNN's discriminative detection power is useful for this topic as a literature review as well. The reason is that automatic pattern detection filters are available instead of designing manual filters in new methods. It is more practical to use CNN's self-learning adaptive filters instead of conventional image processing methods for feature extraction. Moreover, user-friendly interfaces of programing IDE's facilitate easy model creation.

TABLE 2.2 Literature Review Summary.

Study	Method & classifier type	Data description	Accuracy (%)
[34]	PSO + OPF, bootstrap SVM, Pre-selection filter + exhaustive search + SVM	Oxford Parkinson's Disease dataset features plus Pitch Period Entropy (PPE)	Bootstrap with 50 replicates 91.4 ± 4.4%
[32]	Non-linear Dirichlet mixtures, dpMNL, decision trees, SVM	Parkinson's disease voice dataset	87.7 ± 3.3% (five-fold CV)
[27]	OPF, SVM–RBF, SVM–LINEAR, ANN–MLP, SOM, KNN	Speech dataset: MDVP, NHR, HNR, RPDE, D2, DFA, Spread1, Spread2, PPE	75.37 ± 3.58% (random test data)
[33]	Multiclass multi-kernel relevance vector machines (Improved mRVMs)	Oxford Parkinson's Disease dataset	89.55 ± 6.6 % (10-fold CV)
[36]	mRMR + SVM	Oxford Parkinson's Disease Detection dataset: MDVP, NHR, HNR, RPDE, D2, DFA, Spread1, Spread2, PPE	92.8 ± 1.2% (bootstrap with 50 replicates)
[37]	ANN	Parkinson's disease speech dataset	92.9% (65% training and 35% testing)
[38]	GP–EM	Oxford Parkinson's Disease Database (OPDD)	93.1 ± 2.9% (10-fold CV)
[48]	CFS–RF	Parkinson's disease voice dataset: MDVP, Jitter DDP, APQ3	87.1% (10-fold CV)
[47]	Fuzzy-based non-linear transformation + SVM, PCA, KPCA	Different sort of medical datasets and Parkinson's disease voice dataset	93.47% (hold-out)
[46]	Fuzzy entropy measures + Similarity classifier	Parkinson's disease voice dataset	85.03% (hold-out)
[40]	Parallel NN	Oxford Parkinson's Disease voice database	91.20% (hold-out)
[35]	OPF, PSO + OPF, HS + OPF, GSA + OPF	Oxford Parkinson's Disease Detection dataset: MDVP, NHR, HNR, RPDE, D2, DFA, Spread1, Spread2, PPE	PSO + OPF 73.53%, HS + OPF, GSA + OPF 84.01% (hold-out)

TABLE 2.2 (Continued)

Study	Method & classifier type	Data description	Accuracy (%)
[39]	LASSO, mRMR, RELIEF, LLBFS, Random forest, SVM	Parkinson's disease speech dataset (dysphonia measures)	99% (overall value on test data)
[42]	RF ensemble of IBk, SVM	Parkinson's Disease voice signal dataset	97% (test data)
[45]	PSO–FKNN	Parkinson's disease voice dataset from UCI database	97.47% (10-fold CV)
[9]	fuzzy C-means, ANN	Speech signal dataset	fuzzy C-means 68.04% ANN 92 %
[44]	Multinomial logistic regression, Haar wavelets	Parkinson's Disease voice dataset	100% (test data)
[41]	PCA–FKNN	Parkinson's Disease speech dataset of UCI	96.07% (average10-fold CV)
[22]	Gaussian mixture model, PCA, LDA, SFS, SBS, LS-SVM, PNN, GRNN	Voice signals: MDVP, NHR and HNR, RPDE and D2, DFA, Spread1, Spread2, and PPE	100% (test data)
[30]	Decision tree, Ripper k, Bayesian Network	Handwritten drawings: Spiral Average Error, Spiral Cross, Spiral Pressure Ratio, Spiral Side Ratio, etc.	86.67 ± 13.54% (10-fold CV)
[31]	Air movement base data collection, SVM	Handwritten drawings: Online in-air & on-surface movement-based features	85.61% (test data)
[43]	Linear logistic regression, Haar wavelets	Parkinson's Disease voice dataset	100% (test data)
[11]	Color thresholding, blur, median, OPF, NB, SVM	Handwritten drawing dataset: RMS, argmax, argmin, standard deviation, MRT	66.72 ± 5.33% (four-fold CV)
[16]	KNN, ensemble Adaboost, SVM	Parkinson's disease handwritten kinematic features dataset: stroke speed, velocity, acceleration, etc.	82% (test data)
[26]	cifar10, ImageNet, LeNet, OPF	Handwritten drawing: Microphone, Finger grip, Axial Pressure of ink Refill, x, y, z, feature images	85% (25% of test data)

TABLE 2.2 (Continued)

Study	Method & classifier type	Data description	Accuracy (%)
[49]	CNN, NB, SVM	Parkinson's disease handwritten drawings dataset, feature-based images	88.89% (25% test data) 79.64% (10-fold CV)
[21]	Time series-based feature images, CNN: CIFAR10, ImageNet, OPF, NB, SVM	Handwritten drawing: Microphone, Finger grip, Axial Pressure of ink Refill, x, y, z, feature images	95% (voting decision)
[25]	Optimized cuttlefish, KNN, Decision tree	Parkinson Hand, speech, voice datasets	94% (mean value)
[20]	Image skeletonization by web app, OPF, SVM, NB	Collecting handwritten drawing dataset by web app	100% (test data)
[23]	Visual data augmentation, CNN–SVM	Parkinson's disease handwritten dataset: raw, median, edge	83% (voting decision)
[28]	Linking drawing points, SVM linear, SVM–RFB, RF, ET, ADA	Parkinson handwritten drawings: raw image, median filter, edge images	88% (test data)
[29]	sEMG, Optimal topology of ANN and SVM	Handwritten drawings: Density ratio, Height ratio, Execution time, gyroscope components, RMS, etc.	89% (test data)
[24]	Modified GWO, Random Forest, KNN, Decision tree	Handwritten drawing, voice datasets	94% (mean value)

In the literature, the testing stage is generally performed by repeating the test data evaluation and taking the average value. However, cross-validation, leave-one-out cross-validation and voting decision were used as the performance metrics for evaluating the models in some cases. Besides, early stopping condition of the learning process could be implemented for more optimized progression of the training and to obtain the optimum efficiency in the resources. Although many parameters can measure the performance of the machine learning methods, the accuracy metric comes to the forefront in the literature, and this parameter widely is used for performance comparison.

It has been observed that the preprocessing stage have a positive contribution to the accuracy rate. Since there is no specific rule for machine learning, the literature will be enriched if different preprocessing methods are applied. Basically, considering all the studies examined during the research with any preprocessing procedure (about more than 50 works), studies on voice data resulted in an average of 90.5% accuracy; furthermore, studies on handwriting data resulted in an average accuracy of 87.8%. Figure 2.2 is a visual illustration of the average accuracy values of all analyzed works.

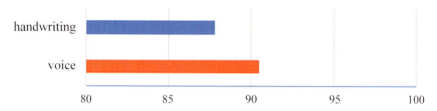

FIGURE 2.2 Average accuracy percentage of studies.

When literature is examined exhaustively, it is seen that higher accuracy percentages are observed in recent studies than the previous studies. One reason for this is that machine learning libraries used in programming languages have become more user-friendly. Another reason is that as time goes on, many different methods have been tried in the literature, more and more optimized methods have emerged accordingly, the researchers have started to make more comprehensive and solution-oriented studies by benefiting from these developments.

In some studies' preprocessing section, different mathematical formulas are developed for novel attributes creation and extraction from raw time-series data. These approaches attempt to increase the discriminative power of the dataset. Raw data often need to be changed through some preprocessing procedure due to the dataset is not suitable for machine learning analysis by default. In the research process, more than 50 studies are examined; consequently, Figure 2.3 shows that the studies in the field of voice data are more than the handwriting ones in general.

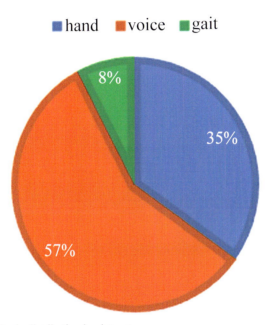

FIGURE 2.3 Study distribution by datasets.

Figure 2.4 illustrates the methods which are used in the PD literature and their use case percentage among all analyzed studies respectively. SVM classifier is generally used for classification on both voice and handwriting data. The figure shows that SVM, NB and OPF classifiers are often used for this literature.

Innovative approaches are included in this study, as well as attributes and methods that have become standard in PD diagnosis for many years. For instance, generally the attributes collected in a handwriting dataset are x, y, z, pressure, grip angle and timestamp. However[51] calculate attributes such as speed, acceleration, RMS, etc. by innovative formulas. Moreover, another example of creativity in the literature is creating unique feature-based images from handwriting raw data.[26] proposed new formulas for feature calculation then transform this feature into images in order to utilize CNN models for classification. Furthermore, as a completely different approach, brain tomography of the patients was used as input data to the image processing and classification models in order to PD diagnosis.[51–56]

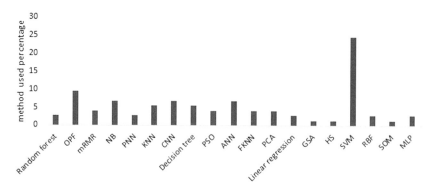

FIGURE 2.4   Study distribution by used methods.

2.3.3   FUTURE RESEARCH DIRECTIONS

This section introduces some useful, practical and theoretical suggestions for potential research fields to researchers. The authors have tried to select and propose novel approaches in order to expand perspective readers in case of PD classification literature.

Considering the successful results of the CNN structures on the PD data classification; 1-D CNN architecture can be implemented as a feature learning model and classifier for signal-based time-series PD raw data.

In order to create a practical and useful dataset, voice and handwriting attributes can be used as a feature combination for each PD patient. This approach requires a tablet device and a microphone for gathering data. Theoretically, this tendency will boost any classification model's performance.

Voice and handwriting data which are collected from patients can be combined with the patients' brain tomography to form a hybrid PD dataset. The main idea is increasing discriminatory power of PD dataset's features and using a voting decision in order to better classification result.

Handling datasets through the preprocessing stage is the main aspect of the work for this literature. Using different optimization algorithms will enrich the literature. Therefore, the comparison of the results before and after preprocessing stage shows the preprocessing stage's performance and importance.

Due to trial and error mentality of machine learning algorithms, comparison of different hybrid models is very useful in order to find out the optimum model. The combinations of different CNN architectures, for instance, cifar10, ImageNet, LeNet, ResNet, VGG16, etc. and highly recommend machine-learning classifiers, for instance, OPF, SVM, NB, KNN, random forest, decision trees, MLP, ANN, SOM, RBF, linear SVM, Ripper k, fuzzy-KNN, fuzzy C-means, etc. can be implemented on PD datasets in order to initiate novel and unique hybrid classification models.

Additionally, when the dataset is in a time-series format, different feature-based images can be extracted from the dataset. In other words, if the dataset is collected as signals, it can be shaped in many forms of information that researchers want. The main purpose of this suggestion is utilizing all the dataset features for the classification process. Therefore, it is highly recommended for collecting the dataset as time-series features.

Moreover, in some articles the author team introduce the PC, mobile, and tablet app for collecting online tests information in order to create a dataset from patients; if medical experts and researchers introduce some tests as standards, then this system can be transformed as an online PD detection system, finally.

In addition, Voice signal's frequencies sketches could be considered as CNN input data in the form of images. Likewise, it will also be useful to create hybrid architectures by using some optimization algorithms like cuttlefish, grey wolf, or ant colony optimization, etc.

Finally, some preprocessing approaches can improve the quality of raw data in order to have an effective classification. For instance, many different filters can be used in preprocessing stage such as salt-pepper filter, blur median, low pass filter, Mean Filter, Gaussian Smoothing, Conservative Smoothing, Crimmins Speckle Removal Frequency Filters, Laplacian/Laplacian of Gaussian Filter, Unsharp Filter.

As seen in Table 2.2, the previous basic datasets have been updated and enriched by developing some new methods which create new features from the features of these datasets. Statistically, analysis of the literature shows that researchers are trying to find the most discriminative feature to obtain better classification results.

2.4 CONCLUSION

This chapter presented a comprehensive review of the prediction of the Parkinson disease by examining the papers that were using machine learning-based approaches. These studies usually attempted to diagnose the PD from voice and handwriting datasets. Essentially, the diagnosis of PD is a classification problem. In the medical diagnostics process, experts always need data and tests to facilitate and support their decisions. Therefore, it is vital to provide auxiliary data for the diagnosis of this disease with machine learning methods.

In this study, the accuracy percentages obtained with the methods that had been used in the related papers were used as the basic criterion to make a comparison between these methods. Additionally, analysis of the literature showed that many different preprocessing methods might lead to obtaining high accuracy results. Since the data were collected in different ways in the handwriting dataset, it required a more detailed preprocessing stage and some better results were obtained. Another result of the analysis of the literature was to point out the importance of using hybrid machine learning, deep learning, and mathematical algorithms. The progress of PD literature showed that new researchers had to find some efficient and practical approaches for the desired automatic PD system.

In conclusion, present models and results are not enough for a fully automatic PD diagnosis system. Accordingly, current methods need a medical expert decision for final diagnosis stage. However, this literature helps in identify early symptoms of PD and encourage doctors for more medical tests to ensure the diagnosis of PD. The literature improves with the new experiments and new ideas, and all works concern is designing the automatic disease detection system.

KEYWORDS

- **Parkinson's disease**
- **machine learning**
- **deep learning**
- **literature review**
- **convolutional neural networks**

REFERENCES

1. Betarbet, R.; Sherer, T. B.; Greenamyre, J. T. J. B. Animal Models of Parkinson's Disease. **2002,** *24* (4), 308–318.
2. Poewe, W.; Seppi, K.; Tanner, C. M.; Halliday, G. M.; Brundin, P.; Volkmann, J.; Schrag, A.-E.; Lang, A. E. J. N. r. D. p. Parkinson Disease. **2017,** *3*, 17013.
3. Beal, M. F. J. N. R. N. Experimental Models of Parkinson's Disease. **2001,** *2* (5), 325.
4. Wooten, G.; Currie, L.; Bovbjerg, V.; Lee, J.; Patrie, J. J. J. o. N. Neurosurgery; Psychiatry, Are Men at Greater Risk for Parkinson's Disease Than Women? **2004,** *75* (4), 637–639.
5. Manciocco, A.; Chiarotti, F.; Vitale, A.; Calamandrei, G.; Laviola, G.; Alleva, E. J. N.; Reviews, B. The Application of Russell and Burch 3R Principle in Rodent Models of Neurodegenerative Disease: The Case of Parkinson's Disease. **2009,** *33* (1), 18–32.
6. Asad, M. N.; Cantürk, İ.; Genç, F. Z.; Özyılmaz, L. In *Investigation of Bone Age Assessment with Convolutional Neural Network by Using DoG Filtering and à Trous Wavelet as Preprocessing Techniques*, 2018 6th International Conference on Control Engineering & Information Technology (CEIT); IEEE, 2018; pp 1–7.
7. Ho, A. K.; Iansek, R.; Marigliani, C.; Bradshaw, J. L.; Gates, S. J. B. n. Speech Impairment in a Large Sample of Patients with Parkinson's Disease. **1999,** *11* (3), 131–137.
8. Cho, C.-W.; Chao, W.-H.; Lin, S.-H.; Chen, Y.-Y. J. E. S. w. a. A Vision-based Analysis System for Gait Recognition in Patients with Parkinson's Disease. **2009,** *36* (3), 7033–7039.
9. Rustempasic, I.; Can, M. J. S. E. J. o. S. C. Diagnosis of Parkinson's Disease Using Fuzzy c-means Clustering and Pattern Recognition. **2013,** *2* (1).
10. Rosenblum, S.; Samuel, M.; Zlotnik, S.; Erikh, I.; Schlesinger, I. J. J. o. n. Handwriting as an Objective Tool for Parkinson's Disease Diagnosis. **2013,** *260* (9), 2357–2361.
11. Pereira, C. R.; Pereira, D. R.; Silva, F. A.; Masieiro, J. P.; Weber, S. A.; Hook, C.; Papa, J. P. J. C. m.; Biomedicine, p. i. A New Computer Vision-based Approach to Aid the Diagnosis of Parkinson's Disease. **2016,** *136*, 79–88.
12. Tseng, M. H.; Cermak, S. A. J. A. J. o. O. T. The Influence of Ergonomic Factors and Perceptual–motor Abilities on Handwriting Performance. **1993,** *47* (10), 919–926.
13. De Stefano, C.; Fontanella, F.; Impedovo, D.; Pirlo, G.; di Freca, A. S. J. P. R. L. Handwriting Analysis to Support Neurodegenerative Diseases Diagnosis: A Review. **2019,** *121*, 37–45.
14. Impedovo, D.; Pirlo, G. J. I. r. i. b. e. Dynamic Handwriting Analysis for the Assessment of Neurodegenerative Diseases: A Pattern Recognition Perspective. **2018,** *12*, 209–220.
15. Drotár, P.; Mekyska, J.; Rektorová, I.; Masarová, L.; Smékal, Z.; Faundez-Zanuy, M. J. I. T. o. N. S.; Engineering, R. Decision Support Framework for Parkinson's Disease Based on Novel Handwriting Markers. **2014,** *23* (3), 508–516.
16. Drotár, P.; Mekyska, J.; Rektorová, I.; Masarová, L.; Smékal, Z.; Faundez-Zanuy, M. J. A. i. i. M. Evaluation of Handwriting Kinematics and Pressure for Differential Diagnosis of Parkinson's Disease. **2016,** *67*, 39–46.
17. Diaz-Cabrera, M.; Gomez-Barrero, M.; Morales, A.; Ferrer, M. A.; Galbally, J. In *Generation of Enhanced Synthetic Off-line Signatures Based on Real On-line Data*,

2014 14th International Conference on Frontiers in Handwriting Recognition; IEEE, 2014; pp 482–487.
18. Galbally, J.; Diaz-Cabrera, M.; Ferrer, M. A.; Gomez-Barrero, M.; Morales, A.; Fierrez, J. J. P. R. On-line Signature Recognition through the Combination of Real Dynamic Data and Synthetically Generated Static Data. **2015,** *48* (9), 2921–2934.
19. Singh, N.; Pillay, V.; Choonara, Y. E. J. P. i. n. Advances in the Treatment of Parkinson's Disease. **2007,** *81* (1), 29–44.
20. Bernardo, L. S.; Quezada, A.; Munoz, R.; Maia, F. M.; Pereira, C. R.; Wu, W.; de Albuquerque, V. H. C. J. P. R. L. Handwritten Pattern Recognition for Early Parkinson's Disease Diagnosis. **2019,** *125*, 78–84.
21. Pereira, C. R.; Pereira, D. R.; Rosa, G. H.; Albuquerque, V. H.; Weber, S. A.; Hook, C.; Papa, J. P. J. A. i. i. m. Handwritten Dynamics Assessment through Convolutional Neural Networks: An Application to Parkinson's Disease Identification. **2018,** *87*, 67–77.
22. Hariharan, M.; Polat, K.; Sindhu, R. J. C. m.; biomedicine, p. i. A New Hybrid Intelligent System for Accurate Detection of Parkinson's Disease. **2014,** *113* (3), 904–913.
23. Moetesum, M.; Siddiqi, I.; Vincent, N.; Cloppet, F. J. P. R. L. Assessing Visual Attributes of Handwriting for Prediction of Neurological Disorders—A Case Study on Parkinson's Disease. **2019,** *121*, 19–27.
24. Sharma, P.; Sundaram, S.; Sharma, M.; Sharma, A.; Gupta, D. J. C. S. R. Diagnosis of Parkinson's Disease Using Modified Grey Wolf Optimization. **2019,** *54*, 100–115.
25. Gupta, D.; Julka, A.; Jain, S.; Aggarwal, T.; Khanna, A.; Arunkumar, N.; de Albuquerque, V. H. C. J. C. s. r. Optimized Cuttlefish Algorithm for Diagnosis of Parkinson's Disease. **2018,** *52*, 36–48.
26. Pereira, C. R.; Weber, S. A.; Hook, C.; Rosa, G. H.; Papa, J. P. In *Deep Learning-aided Parkinson's Disease Diagnosis from Handwritten Dynamics*, 2016 29th SIBGRAPI Conference on Graphics, Patterns and Images (SIBGRAPI); IEEE, 2016; pp 340–346.
27. Spadoto, A. A.; Guido, R. C.; Papa, J. P.; Falcão, A. X. In *Parkinson's Disease Identification through Optimum-path Forest*, 2010 Annual International Conference of the IEEE Engineering in Medicine and Biology; IEEE, 2010; pp 6087–6090.
28. Diaz, M.; Ferrer, M. A.; Impedovo, D.; Pirlo, G.; Vessio, G. J. P. R. L. Dynamically Enhanced Static Handwriting Representation for Parkinson's Disease Detection. **2019,** *128*, 204–210.
29. Loconsole, C.; Cascarano, G. D.; Brunetti, A.; Trotta, G. F.; Losavio, G.; Bevilacqua, V.; Di Sciascio, E. J. P. R. L. A Model-free Technique Based on Computer Vision and sEMG for Classification in Parkinson's Disease by Using Computer-assisted Handwriting Analysis. **2019,** *121*, 28–36.
30. Graça, R.; e Castro, R. S.; Cevada, J. In *Parkdetect: Early Diagnosing Parkinson's Disease*, 2014 IEEE International Symposium on Medical Measurements and Applications (MeMeA). IEEE, 2014; pp 1–6.
31. Drotar, P.; Mekyska, J.; Rektorová, I.; Masarová, L.; Smékal, Z.; Faundez-Zanuy, M. J. C. m.; Biomedicine, p. i. Analysis of In-air Movement in Handwriting: A Novel Marker for Parkinson's Disease. **2014,** *117* (3), 405–411.
32. Shahbaba, B.; Neal, R. J. J. o. M. L. R. Nonlinear Models Using Dirichlet Process Mixtures. **2009,** *10*, 1829–1850.

33. Psorakis, I.; Damoulas, T.; Girolami, M. A. J. I. T. o. n. n. Multiclass Relevance Vector Machines: Sparsity and Accuracy. **2010,** *21* (10), 1588–1598.
34. Little, M.; McSharry, P.; Hunter, E.; Spielman, J.; Ramig, L. J. N. P. Suitability of Dysphonia Measurements for Telemonitoring of Parkinson's Disease. **2008,** 1–1.
35. Spadoto, A. A.; Guido, R. C.; Carnevali, F. L.; Pagnin, A. F.; Falcão, A. X.; Papa, J. P. In *Improving Parkinson's Disease Identification through Evolutionary-based Feature Selection*, 2011 Annual International Conference of the IEEE Engineering in Medicine and Biology Society; IEEE, 2011; pp 7857–7860.
36. Sakar, C. O.; Kursun, O. J. J. o. m. s. Telediagnosis of Parkinson's Disease Using Measurements of Dysphonia. **2010,** *34* (4), 591–599.
37. Das, R. J. E. S. w. A. A Comparison of Multiple Classification Methods for Diagnosis of Parkinson Disease. **2010,** *37* (2), 1568–1572.
38. Guo, P.-F.; Bhattacharya, P.; Kharma, N. In *Advances in Detecting Parkinson's Disease*, International Conference on Medical Biometrics; Springer, 2010; pp 306–314.
39. Tsanas, A.; Little, M. A.; McSharry, P. E.; Spielman, J.; Ramig, L. O. J. I. t. o. b. e. Novel Speech Signal Processing Algorithms for High-accuracy Classification of Parkinson's Disease. **2012,** *59* (5), 1264–1271.
40. Åström, F.; Koker, R. J. E. s. w. a. A Parallel Neural Network Approach to Prediction of Parkinson's Disease. **2011,** *38* (10), 12470–12474.
41. Chen, H.-L.; Huang, C.-C.; Yu, X.-G.; Xu, X.; Sun, X.; Wang, G.; Wang, S.-J. J. E. s. w. a. An Efficient Diagnosis System for Detection of Parkinson's Disease Using Fuzzy k-nearest Neighbor Approach. **2013,** *40* (1), 263–271.
42. Ozcift, A. J. J. o. m. s. SVM Feature Selection Based Rotation Forest Ensemble Classifiers to Improve Computer-aided Diagnosis of Parkinson Disease. **2012,** *36* (4), 2141–2147.
43. Mandal, I.; Sairam, N. J. I. J. o. S. S. New Machine-learning Algorithms for Prediction of Parkinson's Disease. **2014,** *45* (3), 647–666.
44. Mandal, I.; Sairam, N. J. I. j. o. m. i. Accurate Telemonitoring of Parkinson's Disease Diagnosis Using Robust Inference System. **2013,** *82* (5), 359–377.
45. Zuo, W.-L.; Wang, Z.-Y.; Liu, T.; Chen, H.-L. J. B. S. P.; Control, Effective Detection of Parkinson's Disease Using an Adaptive Fuzzy k-nearest Neighbor Approach. **2013,** *8* (4), 364–373.
46. Luukka, P. J. E. S. w. A. Feature Selection Using Fuzzy Entropy Measures with Similarity Classifier. **2011,** *38* (4), 4600–4607.
47. Li, D.-C.; Liu, C.-W.; Hu, S. C. J. A. I. i. M. A Fuzzy-based Data Transformation for Feature Extraction to Increase Classification Performance with Small Medical Data Sets. **2011,** *52* (1), 45–52.
48. Ozcift, A.; Gulten, A. J. C. m.; Biomedicine, p. i. Classifier Ensemble Construction with Rotation Forest to Improve Medical Diagnosis Performance of Machine Learning Algorithms. **2011,** *104* (3), 443–451.
49. Khatamino, P.; Cantürk, İ.; Özyılmaz, L. In *A Deep Learning-CNN Based System for Medical Diagnosis: An Application on Parkinson's Disease Handwriting Drawings*, 2018 6th International Conference on Control Engineering & Information Technology (CEIT); IEEE, 2018; pp 1–6.

50. Isenkul, M.; Sakar, B.; Kursun, O. In *Improved Spiral Test Using Digitized Graphics Tablet for Monitoring Parkinson's Disease*, Proceedings of the International Conference on e-Health and Telemedicine; 2014; pp 171–175.
51. Illán, I.; Górriz, J.; Ramírez, J.; Segovia, F.; Jiménez-Hoyuela, J.; Ortega Lozano, S. J. M. p. Automatic Assistance to Parkinson's Disease Diagnosis in DaTSCAN SPECT Imaging. **2012,** *39* (10), 5971–5980.
52. Khare, N.; Devan, P.; Chowdhary, C. L.; Bhattacharya, S.; Singh, G.; Singh, S.; Yoon, B. SMO-DNN: Spider Monkey Optimization and Deep Neural Network Hybrid Classifier Model for Intrusion Detection. *Electronics* **2020,** *9* (4), 692.
53. Chowdhary, C. L.; Acharjya, D. P. Segmentation and Feature Extraction in Medical Imaging: A Systematic Review. Procedia Comput. Sci. **2020,** *167,* 26–36.
54. Das, T. K.; Chowdhary, C. L.; Gao, X. Z. Chest X-Ray Investigation: A Convolutional Neural Network Approach. *J. Biomimetics, Biomater. Biomed. Eng.* **2020,** *45,* 57–70.

CHAPTER 3

Machine Learning Algorithms for Hypertensive Retinopathy Detection through Retinal Fundus Images

N. JAGAN MOHAN*, R. MURUGAN, and TRIPTI GOEL

Department of Electronics and Communication Engineering, National Institute of Technology Silchar, Assam 788010, India

*Corresponding author. E-mail: jaganmohan427@gmail.com

ABSTRACT

Hypertensive retinopathy (HR) is a retinal sickness that is caused because of reliably (hypertension) and prompts vision misfortune. A great many individuals on the planet are experiencing HR illness because of hypertension. The variations from the norm happen on the retina because of hypertension. This sickness does not have any early signs and much of the time, HR is analyzed at later stages when the illness prompts visual deficiency or vision misfortune. It is fundamental for hypertensive patients to have a normal assessment of their eyes. This chapter deals with the description of HR, for example, classification, symptoms, and related risk factors. It also deals with the comparative analysis of the algorithms proposed by the researchers on how the machine learning approaches are more accurate in automatic detection of HR, for example, conventional methods and machine learning approaches proposed for the detection of HR.

3.1 INTRODUCTION

The human eye is an incredibly perplexing structure that empowers locate, one of the most significant of the human detects. Sight underlies our capacity to comprehend our general surroundings and to explore inside our condition. As we take a gander at our general surroundings, our eyes are always taking in light, a part basic to the visual procedure. The retina is a layer of sensory tissue that covers within the back 66% of the eyeball, in which incitement by light happens, starting the impression of vision. The retina is really an augmentation of the cerebrum, shaped embryonically from neural tissue and associated with the cerebrum legitimate by the optic nerve. The retina capacities explicitly to get light and to change over it into substance vitality.[2,8] The concoction vitality initiates nerves that direct the electrical messages out of the eye into the higher districts of the mind.

> *Smart animals have dumb retinas and*
> *dumb animals have smart retinas.*

Hypertensive retinopathy (HR) is the medicinal term brought about by hypertension. It, for the most part, influences the retina and retinal blood dissemination. Because of hypertension, retinal blood vessels (BV), for example, retinal corridors and retinal veins are likewise influenced. So blood dissemination to the retina crumbled. The indication of HR relies upon the patient's conditions.[17,26] Some ailments may have visual conditions. HR side effects incorporate vein changes, supply route narrowing, vein narrowing, conduit vein crossing area edge deviation; this is known as arteriovenous scratching. This arteriovenous scratching, for the most part, influences the supply routes and veins crossing areas.[17,18]

This chapter deals with the detailed description of the HR, for example, classification or grading, symptoms and risk factors related to HR. This book chapter deals with the comparative analysis of the algorithms proposed by the researchers or scientists how the machine learning (ML) approaches are more accurate in automatic detection of HR, for example, conventional methods and ML approaches proposed for the detection of HR.

3.1.1 THE EYE FUNDUS

The advanced field of ophthalmology was borne from hundreds of years of perception and revelation that inevitably became grounded in logical information. A noteworthy advance in the comprehension and finding of eye ailments was the improvement in the nineteenth century of the ophthalmoscope, an instrument for investigating the inside of the eye. With this gadget, ophthalmologists could promptly inspect the retina and its veins, in this manner acquiring important data about the inward eye and eye maladies[15].

The ophthalmoscope instrument is used for reviewing the inside of the eye. It was created in 1850 by German researcher, furthermore, savant Hermann von Helmholtz. The ophthalmoscope turned into a model for later types of endoscopy. The gadget comprises of a solid light that can be coordinated into the eye by a little mirror or crystal. The light reflects off the retina and back through a little opening in the ophthalmoscope, through which the inspector sees a non-stereoscopic amplified picture of the structures at the back of the eye, including the optic plate, retina, retinal BV's, and macula shown in Figure 3.1. The ophthalmoscope is especially helpful as a screening device for different visual infections, for example, diabetic retinopathy (DR).[15]

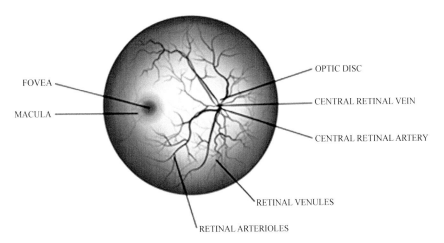

FIGURE 3.1 The retinal fundus image.

Source: Reprinted with permission from Carolina Ophthalmology, open access.

The appropriate upkeep of the retina is exceptionally essential for good vision. There are different eye-related infections like DR, HR, Retinopathy of Rashness, and Retinal Vein Occlusion which, for the most part, influences the retina. In the event that they stay undiscovered for quite a while, it can lead to loss of vision. Specialists and doctors recognize these retinal sicknesses when signs like hemorrhages (HEM), delicate and hard Exudates (EX), optic plate growing and arteriolar narrowing are present. There are different parameters utilizing which they can grade the seriousness of retinal sicknesses[25] (Fig. 3.2).

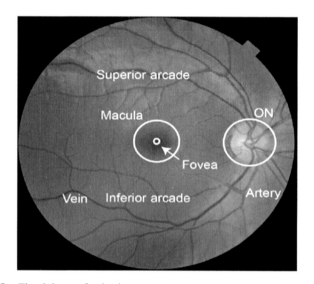

FIGURE 3.2 The right eye fundus image.
Source: Reprinted with permission from Ref. [11] © Elsevier.

The focal point of the fundus lies on the optical hub; this is the fovea, which gathers the best-settled pictures, and it is typically connected with a little yellow dab, the macula lutea. The anatomic and clinical foveola, fovea, and macula are shown on the outline. The major vascular inventory of the retina structures from the predominant and substandard arcade of veins. The retinal region between the predominant and substandard arcade is known as the territory central or back post. The focal point of this back post contains the macula, which is redder (dim dark in the print adaptation) and denser in shading than the encompassing retina. This is because of more photoreceptors stuffed at high densities what's more, more colors behind the photoreceptor cells. The macula lutea alludes to yellow xanthophyll color inside the retina in the focal point of the macula. The focal point of the macula is alluded to as the fovea, which is 500 µm in distance across an avascular region that basically made out of the internal restricting layer and concentrated cone photoreceptor cells, known as the pack of Rochelle Duverney. The significant vessels unmistakable in this shading fundus photographs lie in the superior retina.[11,31]

3.2 HYPERTENSIVE RETINOPATHY

The retina is an inside and significant piece of the human eye whose work is to catch and send pictures to the cerebrum. It comprises of various structures alongside two sorts of veins, veins and courses. These retinal veins are influenced by the quantity of eyes maladies. HR is caused because of a steady high pulse in retinal BV's. A great deal of people groups on the planet is experiencing HR sickness; be that as it may, by and large, HR patients are ignorant of it. The presence of HR and its seriousness can be distinguished by the patient's eye ophthalmologic assessment. More often than not, HR is analyzed at the last stage which drove the patient to visual deficiency or vision misfortune; thusly, it is important for HR patients to ensure the standard assessment of their eyes.[3] Presumably, there will not be any signs until the condition has advanced broadly. Potential signs and indications include: compact visualization, eye-distension, teeming of a BV, double visualization accompanied by headaches. Getting quick restorative assistance is better if the circulatory strain is high; and all of a sudden has changed in the vision.[9]

Clinical discoveries of HR incorporate the presence of sores which can be ordered into two gatherings, for example, delicate exudates and hard exudates. Delicate exudates are otherwise called Cotton Wool Spots (CWS). CWS is soft white-yellow spots seen in cutting edge phases of HR, though HE is splendid yellow injuries. These CWS are either observed detached in fundus pictures or exist with different injuries like HEM and HE of tissue's blood supply. CWS is likewise found in the retina of diabetic patients yet they are all the more firmly identified with HR when contrasted with DR. DR is described by different HE and a couple of CWS while numerous CWS is related to HR.[14] This sickness does not have early ciphers and much of the time, HR is examined at later phases when the illness prompts visual deficiency or vision misfortune. In this way, it is fundamental for hypertensive patients to have a normal assessment of their eyes.

Drawn out hypertension, or hypertension is the primary driver of HR. Hypertension is an interminable issue where the blood power against the arteries is excessively high. The power is an after effect of the blood siphoning out from the heart and to the supply routes just as the power made as the heart rests between pulses. When the blood travels through the BV with more pressure, in the long run it makes harm by extending the supply routes. This prompts numerous issues after some time. HR, for the most part, happens after the pulse has been reliably high over a drawn-out period. The blood pressure (BP) can be influenced by: not having a daily physical activity, being fatty, taking too much salt in daily food, daily stress, High BP.[6]

HR is analyzed dependent on its clinical appearance on the widened funduscopic test and concurrent hypertension. The primary care physician will utilize an ophthalmoscope to analyze the retina. It sparkles a light through the understudy for inspecting the rear of the eye for indications of tightening veins or to check whether any liquid is spilling from the veins. This strategy is easy. It takes under 10 min to finish. At times, a unique test called fluorescein angiography (FA) is performed to look at the retinal bloodstream. In this strategy, the doctor applies distinct eye droplets to enlarge the pupils and afterward takes the photos of the eye. After taking the pictures, the primary care physician will infuse a colorant called fluorescein into a vein. They will commonly do this within the elbow. As the dye moves into the veins of the eye the retinal images are accepted. Intense harmful hypertension will make patients grumble of eye agony, migraines, or diminished visual keenness. Ceaseless arteriosclerotic changes from hypertension would not cause any side effects alone. In any case, the complexities of arteriosclerotic hypertensive changes make patients the present with normal indications of vascular impediments or micro aneurysms (MA). For any disease, it is better to know the severity of that disease so that the required treatments and precautions for the further developments of the disease can be taken. In this purpose, it is required to grade the disease. The following section discusses the grading and classification of the HR.[21]

3.2.1 GRADING HR

Keith and colleagues developed the first classification system for HR in 1939. Since then, the original model has been criticized for the reproducibility and validity of the method in clinical practice.[37] Others claim that the levels of retinopathy may not equate with the extent of systemic hypertension, like Hayreh. Some suggested, however, that classifications could be associated with heart disease. In particular, recent work connects the revised Keith–Wagener–Barker model defined by Mitchell and Wong to the target damage of the end-organ.[10] The following are the

HR classification systems based on the retinal fundus image consideration with the help of ophthalmoscope as indicated in Tables 3.1 and–3.3.

3.2.1.1 KEITH–WAGENER–BARKER CLASSIFICATION (1939)

Based on their ophthalmoscopic findings, patients were grouped. This was, therefore, the first method to associate retinal results with the state of the hypertensive disease. The rankings are as follows:

TABLE 3.1 The Keith–Wagener–Barker HR Classification Based on the Retinal Fundus Image Severity.

Grade	Classification	Symptoms
I-mild hypertension	Gentle summed up retinal arteriolar narrowing or sclerosis	No symptoms
II-more marked HR	Unmistakable central narrowing and arteriovenous intersections. Moderate to checked sclerosis of the retinal arterioles. Misrepresented blood vessel light reflex	Asymptomatic
III-mild angiosplastic retinopathy	Retinal HEM, EX, and CWS. Sclerosis and spastic lesions of retinal arterioles	Indicative
IV	Grade III + papilloedema (Severe)	Compact existence

Source: Adapted from Ref. [10].

TABLE 3.2 Mitchell–Wong Classification of HR.

Grade	Classification
I-mild retinopathy	Arteriolar narrowing, AV scratching as well as arteriolar divider mistiness
II-moderate retinopathy	HEM, MA, CWS, and/or rough EX
III-malignant retinopathy	Grade-II + optical disc-(OD) swelling

Source: Adapted from Ref. [10].

The HR severity is measured in terms of the arteries and veins ratio generally known as AVR ratio. The AVR ratio is calculated using Central Retinal Arterial Equivalent (CRAE) and Central Retina Venous Equivalent (CRVE) measurements. These measurements are described by formulas of Parr–Hubbar.[27] The "Arteriole" and "Venule" contain the mean widths of arteries and veins segments under Region of Interest.

TABLE 3.3 Scheie Classification of HR.

Grade	Scheie classification	The Scheie classification based on light reflex changes from arteriolosclerotic changes	Modified Scheie classification
0	Determination of hypertension, however, no noticeable retinal variations from the norm	Normal	No changes
1	Verbose arteriolar narrowing; no central choking	Expanding of light reflex with negligible arteriovenous pressure	Barely detectable arterial narrowing
2	Increasingly articulated arteriolar narrowing with central choking	Light reflex changes and intersection changes progressively unmistakable	Obvious arterial-narrowing with focal irregularities
3	Central and diffuse narrowing with retinal HEM	Copper-wire appearance; progressively unmistakable arteriovenous pressure	Grade-2 + retinal HEM and/or EX
4	Retinal edema, hard EX, OD edema	Silver-wire appearance; severe arteriolovenous crossing changes	NA

Source: Adapted from Ref. [40].

$$\text{CRAE} = \sqrt[2]{(0.87W_a^2 + 1.01W_b^2 - 0.22W_aW_b - 10.73}$$

Where Wb = median value of "Arteriole" and Wa = the value in the same list exactly before the median.

$$\text{CRVE} = \sqrt[2]{(0.72W_a^2 + 0.91W_b^2 + 450.02}$$

As shown in the Figure 3.3, if the AVR ratio is in between 0.667 and 0.75 then that retinal image is graded as normal. If the AVR ratio is 0.5, 0.33, 0.25, and <0.20 then retinal images are graded as Grade-1, Grade-2, Grade-3, Grade-4, respectively.[39]

The differential diagnosis for HR with diffuse retinal HEM, CWS, and hard EX incorporates most outstandingly DR. DR can be recognized from HR by assessment for the individual fundamental diseases. Different conditions with diffuse retinal HEM that can take after HR incorporate radiation retinopathy, paleness, and other blood dyscrasias, visual ischemic disorder, and retinal vein impediment.

The gradation and extent of hypertension are usually the key determinants of retinopathy with hypertension. The improvements mentioned in the sections above, however, are not specific for hypertension. Similar improvements can be seen in many vascular-risk disorders, such as diabetes. Often, when diabetes and hypertension are involved, retinopathy may be more extreme and progressive. Certain causes, such as hyperlipidemia, may also aggravate retinopathy.[6]

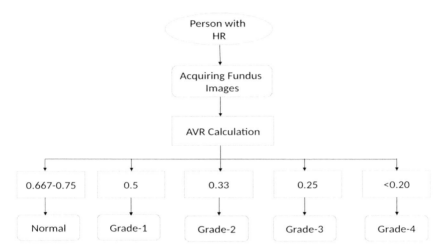

FIGURE 3.3 Grading of HR.
Source: Reprinted with permission from Ref. [39].

Figure 3.4 shows the HR grades of typical advanced retinal fundus photos of mild (a, b), moderate (c, d), and malignant (e, f) HR, as reviewed with the improved characterization. (a) Mild-HR is demonstrated by the nearness of summed up arteriolar narrowing, arteriovenous scratching, and opacification of the arteriolar divider ("copper wiring"). (b) Mild-HR with central arteriolar narrowing. (c and d) Moderate-HR with various retinal HEM and CWS. (e and f) Malignant-HR with the growing of the OD, retinal HEM, hard EX, and CWS.

3.3 PERFORMANCE METRICS

In HR, retinal images division execution measures are condensed, Accuracy (Acc), Sensitivity (Se), and Specificity (Sp) are the most often received measures.

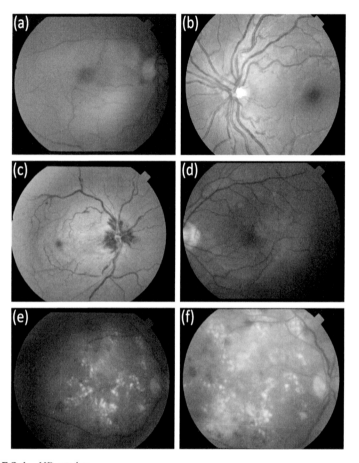

FIGURE 3.4 HR grades.

Source: Reprinted with permission from public database STARE, Open access https://cecas.clemson.edu/~ahoover/stare/.

The retinal vessel order depends on the accurately ordered vessel (TP-True positive) and non-vessel (TN-True negative), and mistakenly grouped vessel (FP-False positive) also, non-vessel (FN-False negative). TP distinguishes that pixel is a vessel in both the sectioned and ground truth picture; while in TN, the pixel is non-vessel in the sectioned and ground truth pictures. FP recognizes that pixel is a vessel in the fragmented picture yet non-vessel in eyewitness stamped picture, additionally in FN, the pixel is a vessel in ground truth while non-vessel in the sectioned picture. These terms are utilized to assess execution.

3.3.1 ACCURACY

Acc is defined as the ratio of correctly identified pixels to the total number of pixels present in the image.

$$\text{Acc} = \frac{TP + FN}{TP + TN + FP + FN} \tag{3.1}$$

3.3.2 SENSITIVITY AND SPECIFICITY

Se and Sp are the factual proportions of the execution of a twofold grouping test in HR. As conveyed in eq 3.2, Se likewise alluded as TP rate, measures the extent of positives, both TP and FN, that are effectively recognized.

As communicated in eq 3.3, Sp estimates the extent of negatives, both TN and FP that are accurately distinguished. Despite the fact that a high Se mirrors the attractive calculation tendency to recognize vessels, a high Se with low Sp shows that the division incorporates numerous pixels that do not have a place with vessels, for example, high FP.

$$\text{Se} = \frac{TP}{TP + FN} \tag{3.2}$$

$$\text{Sp} = \frac{TN}{TN + FP} \tag{3.3}$$

3.3.3 POSITIVE PREDICTED VALUE

It is the capacity measure that the BV pixel identified as the BVs is really positive and it is expressed in eq 3.4.

$$\text{PPV} = \frac{TP}{TP + FP} \tag{3.4}$$

3.3.4 AREA UNDER CURVE

The ratio between the true positive rate and false positive rate is considered as AUC and it is indicated in eq 3.5.

$$\text{AUC} = \frac{1}{2}\left(\frac{TP}{TP + FN} + \frac{TN}{TN + FP}\right) \tag{3.5}$$

3.4 METHODS

There are several methods or techniques available for HR grading. They are broadly classified into two types, that is, conventional approaches and the ML based approaches. The basic idea of these two approaches is presented in the following sections.

3.4.1 CONVENTIONAL METHOD

Fundus photography includes capturing the back of an eye, otherwise called the fundus. Specific fundus cameras comprising of a mind-boggling magnifying instrument appended to a flash permitted camera are utilized in fundus taking photographs. The primary structures that can be envisioned on a fundus are the focal and fringe retina, OD and macula. The general approach for HR grading using conventional approaches is shown in Figure 3.5.

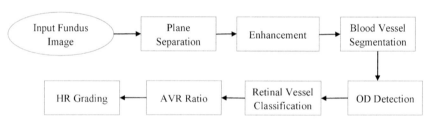

FIGURE 3.5 Conventional approach for HR grading.

Fundus picture is a RGB shading picture, when all is said in done, RGB pictures comprise of three channels (red-green-blue) This can be sophisticated by detachment the retina picture to three channels and utilizing just one of them (Green channel), the blue channel is portrayed by low differentiation and doesn't contain a lot of data. The vessels are obvious in the red channel.[23]

The information picture is resized and the Red or Green channel picture is isolated as the vein seems more brilliant in the Red[3] or on the other hand green channel picture. At that point, the morphological activity is performed on the Red or green channel picture. The essential morphological activities are dilation and erosion. The more unpredictable morphological activities are opening and closing. Dilation is an activity that develops or thickens questions in a binary picture. The particular way and degree of this thickening are constrained by shape alluded to as an organizing component. Dilation is characterized as far as a set activity. Erosion shrivels or diminishes questions in a parallel picture. The way and degree of contracting is constrained by a structured component.

The next step included is image enhancement. Image enhancement strategies are numerical methods that are planned for acknowledging the improvement in the nature of a given picture. The outcome is another picture that shows certain highlights in a way that is better in some sense when contrasted with their appearance in the first picture. One may likewise determine or register different handled forms of the first picture, each introducing a chose to highlight in an upgraded appearance. Straightforward picture improvement systems are created and applied in an impromptu way. Propelled systems that are advanced concerning certain particular prerequisites and target criteria are likewise accessible.

Some of the image enhancement techniques are as mentioned below:

1. Filtering with morphological operators.
2. Histogram equalization.
3. Noise removal using a Wiener filter.
4. Linear contrast adjustment.
5. Median filtering.
6. Unsharp mask filtering.
7. Contrast limited adaptive histogram equalization (CLAHE)
8. Decorrelation stretch.

Adaptive histogram equalization (AHE) is a PC picture getting ready method used to improve separate in pictures. It changes from customary histogram alteration in the respect that the flexible procedure enrolls a couple of histograms, each identifying with an unquestionable section of the image, and uses them to redistribute the delicacy estimations of the image. It is in this manner fitting for improving the area to separate and redesigning the implications of edges in each region of an image. Regardless, AHE tends to over amplify commotion in by and large homogeneous zones of an image. A variety of flexible histogram equalization called CLAHE maintains a strategic distance from this by limiting the improvement.

The BVs are the essential anatomical structure that can be unmistakable in retinal pictures. The division of retinal veins has been acknowledged worldwide for the conclusion of both cardiovascular (CVD) and retinal infections. In this manner, it requires a fitting vessel division technique for the programmed discovery of retinal ailments, for example, diabetic retinopathy and waterfall. The identification of retinal infections utilizing PC supported conclusions can help individuals to keep away from the dangers of visual disability furthermore, spare restorative assets.[33]

The preparing of the retinal fundus picture is the starter step for the vessel division task. It includes various advances, for example catching a photograph of the eye containing vessel, vessel upgrade, expelling commotion and assessing the exhibition utilizing various measures, and so forth.

Utilizing typical division procedures, we can distinguish just the veins. Along these lines, shading picture division is the most ideal approach to distinguish the retinal issues, since utilizing shading picture division we can separate the retinal veins and arteries.

Color fundus picture vein extraction is principally continued to separate the arteries and veins. Retinal vascular organize extraction can be completed by utilizing exceptionally high goals fundus color pictures. It has a few inconveniences, for example, an impression of focal light, antiquity present in the information retinal picture the proposed framework for shading retinal BV division comprises of a mix of morphological procedures to recognize veins.

Some of the BV segmentation methods[22] are shown in Figure 3.6.

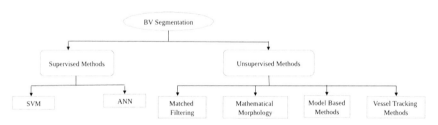

FIGURE 3.6 BV Segmentation methods.
Source: Reprinted with permission from Ref. [22]. © 2020 Springer.

The next step involved is optic disc detection. For retinal images, the optic disk is a central anatomical structure. The ability to detect optical disks for retinal images plays a major role in automatic screening systems. The next step followed by the OD detection is retinal vessel classification. Based on the thresholding values carried out by the pixels retinal vessels are classified into arteries or veins. As mentioned in the above sections the AVR ratio is calculated based on which the HR grading will be done.

3.4.2 MACHINE LEARNING METHODS

Classification of retinal fundus images has become one of the main uses of the pilot to illustrate ML. Convolutional neural networks (CNNs) are a kind of deep neural networks (DNN) that generate fairly accurate results when used to classify retinal fundus images.[29,35-36,42-45] The general approach for grading HR using ML is shown in Figure 3.7.

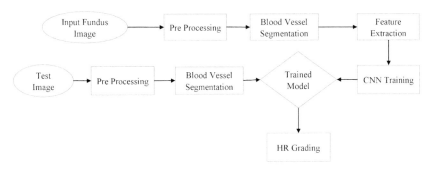

FIGURE 3.7 Machine learning approach for HR grading.

Retinal pictures incorporate pertinent data to HR just as uproarious and unimportant pixels. The evacuation of undesirable pixels is called preprocessing.

The preprocessing of the picture is the key idea for the better division procedure of the retinal vessel before the classifier preparing and testing stage. This progression may include various advances and systems relying upon the necessities of the classifier. This is normally performed for clamor decrease, vessel improvement, and exception cancellation, and so forth.

Some of the preprocessing techniques are as follows:

i. Transforming an image
ii. Extracting green channel image
iii. Morphological operations
iv. Knowledge base processing
v. Enhancing the image
vi. Filtering
vii. BV segmentation
viii. Extracting features
ix. Selecting features
x. Restoring the images

The steps involved next to preprocessing are mention in the earlier topics. After selecting the desired features from the preprocessed retinal fundus image the CNN model can be trained.

A CNN is a particular kind of counterfeit neural system that utilizations perceptrons, an AI unit calculation, for regulated learning, to dissect information. CNN's apply to picture handling, regular language preparing and different sorts of psychological errands. A CNN is otherwise called a "ConvNet." Like different sorts of counterfeit neural systems, a CNN has an info layer, a yield layer, and different concealed layers. A portion of these layers is convolutional, utilizing a numerical model to give results to progressive layers. This recreates a portion of the activities in the human visual cortex. CNNs are an essential case of profound realizing, where an increasingly refined model pushes the advancement of computerized reasoning by offering frameworks that reenact various sorts of natural human mind action. The components of a CNN are as follows:

3.4.2.1 CONVOLUTIONAL LAYER

Convolution layer is an important layer to extract the features from the given image which is carrying information. Convolution keeps the relationship between the pixels by extracting features from the image with the help of square matrices of the given data. It is a scientific activity that takes two information sources, for example, pixels arranged in a matrix and a kernel or part. Based on the filter, we apply on an image, we can find out the borders and we can increase or decrease the quality of the image with certain operations.

The convolution of the pixels values of an image multiplied with filter matrix is called "Feature Map." Let an image matrix is having a dimension of $h \times w \times d$ and the filter is $f_h \times f_w \times d$ then the output dimensions of the image are $(h - f_h + 1) \times (w - f_w + 1) \times 1$. The matrix of the image multiplied with the filter is shown in Figure 3.8.

3.4.2.2 STRIDE

The number of pixels that moves over the given input image matrix is called as stride. For example, if the stride is 1 then we move the kernel to 1 pixel and if the stride is 2 then we move the kernel to 2 pixels on the given input image matrix.

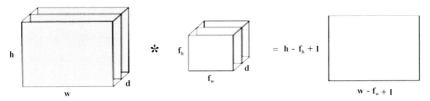

FIGURE 3.8 Image pixels matrix multiplied with kernel or filter matrix.
Source: Reprinted from Ref. [3]. Open access.

3.4.2.3 PADDING

In certain situations, the kernel may not be fitted with the given image pixels matrix. In such situations, we are having two choices

i. We can add zeros to the input image matrix so that the kernel or filter fits
ii. We can drop or eliminate a part of input image where the kernel or filter fits

Rectified Linear Unit for a non-linear operation-ReLU.

Sometimes, there may be a chance to have negative values in the given matrices. To provide non-linearity in ConvNet the ReLU operations will be useful in providing non-negative linear values.

The output of the ReLU is $f(x) = max(0, x)$

3.4.2.4 POOLING

The principal mystery ingredient that has made CNNs exceptionally successful is pooling. Pooling is a vector to scalar change that works on every nearby area of a picture, much the same as convolutions do, be that as it may, in contrast to convolutions, they do not have channels and do not figure dab items with the neighborhood locale, rather, they process the normal of the pixels in the district (Average Pooling) or just picks the pixel with the most noteworthy power and disposes of the rest (Max Pooling).

3.4.2.5 DROPOUTS

Overfitting is a marvel whereby a system functions admirably on the preparation set yet performs inadequately on the test set. This is frequently because of inordinate reliance on the nearness of explicit highlights in the preparation set. Dropout is a strategy for battling over-fitting. It works by haphazardly setting a few initiations to 0, basically executing them. By doing this, the system is compelled to investigate more methods for arranging the pictures rather than over-contingent upon certain highlights. This was one of the key components in the AlexNet.

3.4.2.6 BATCH NORMALIZATION

A significant issue with neural systems is evaporating gradients. This is a circumstance whereby the inclinations become excessively little, consequently, preparing surfers frightfully. Ioffe and Szegedy from Google Brain found this was to a great extent because of the inside covariate move, a circumstance that emerges from the change information appropriation as data moves through the system. What they did was to gadget the method known as bunch standardization. This works by normalizing each group of the picture to have zero mean and unit difference. It is generally set before non-linearity (relu) in CNNs. It significantly improves exactness while fantastically accelerating the preparation procedure.

3.4.2.7 DATA AUGMENTATION

The last fixing required or present-day covnets is information increase. The human vision framework is amazing at adjusting to picture interpretations, pivots, and different types of mutilations. Take a picture and flip it, at any rate, a great many people can, in any case, remember it. In any case, covnets are not truly adept at taking care of such contortions; they could bomb frightfully because of minor interpretations. The way to settling this is to haphazardly misshape the preparation pictures, utilizing flat flipping, vertical flipping, pivot, brightening, moving, and different twists. This would empower covnets to figure out how to deal with these contortions, henceforth, they would have the option to function admirably in reality. Another basic method is to subtract the mean picture from each picture and furthermore isolate it by the standard deviation.

3.4.2.8 FULLY CONNECTED LAYER

It is also known as the FC layer. Matrix has to be leveled into a vector and feed it into a FC layer like a neural system. In this, the element map framework will be altered over as a vector.

With the FC layers, we can join the highlights together to make a model. At long last, we have an initiation capacity, for example, softmax or sigmoid to arrange the yields into various classes. The neural networks with multiple convolutional layers and the complete CNN architectures are as shown in the Figures 3.9 and 3.10, respectively.

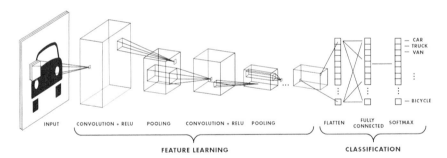

FIGURE 3.9 Neural network with more than one convolutional layers.
Source: Reprinted from Ref. [3]. Open access.

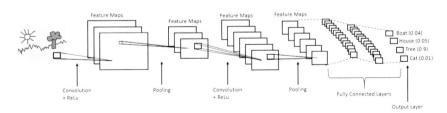

FIGURE 3.10 Complete CNN architecture.
Source: Reprinted from Ref. [3]. Open access.

3.5 DATABASE

There are various open retinal datasets accessible with BV subtleties. It is the key advance for the vein division to prepare and test the classifier on the retinal database. A few databases, for example, DRIVE and STARE and so on are publically accessible for the specialists alongside the ground truth pictures of the vessels. The exhibition of the classifier can be assessed utilizing these datasets.

3.5.1 DIGITAL RETINAL IMAGES FOR VESSEL EXTRACTION-DRIVE

DRIVE is one of the normally utilized datasets for retinal BV division.[24] DRIVE comprises of 40 retinal pictures in which 33 are more beneficial pictures while 7 have given indications of gentle diabetic retinopathy. Group CR5 non-mydriatic camera with 45° field of vi (FOV) and eight pieces for every shading channel at 768 × 584 pixels have been utilized to catch the pictures in the JPEG position. Each picture has a round FOV with 540 pixels' distance across. DRIVE dataset has been isolated into preparing and test set with 20 pictures each. In the preparation set, 14 pictures were fragmented by the first master and 6 pictures were divided constantly. In the test set, the division has been performed twice in two cases. In case 1, first and the second master divided 13 and 7 pictures individually; while the case 2 has been performed by the third master. In case 1 and case 2, the spectators stamped 12.7 and 12.3% pixels as vessels individually.

3.5.2 STRUCTURED ANALYSIS OF THE RETINA-STARE

This dataset comprises of 400 retinal pictures, caught utilizing TOPCON TRV-50 fundus camera with extra settings of 35° FOV and 8 bits/shading channel at 605 × 700 pixels. The normal width of the FOV is 650 × 700. Gaze has 20 vessel ground truth pictures utilized for vein division in which 9 are more beneficial while the rest of them have indicated various kinds of retinal maladies.[41] Two specialists have physically sectioned these pictures where the main master portioned 10.4% vessel pixel, while the subsequent master sectioned 14.9% of the more slender vessel. By and large, the division of the main spectator used to figure the exhibition as the ground truth.[38]

3.5.3 ANNOTATED DATASET FOR VESSEL SEGMENTATION AND CALCULATION OF ARTERIOVENOUS RATIO-AVRDB

AVRDB is a recently created HR database that will be freely accessible at www.biomisa.org in future for the consideration network. It is having 100 fundus retinal pictures that are caught through TOPCON TRC-NW8 and explained with the assistance of master ophthalmologists from the Armed Forces Institute of Ophthalmology. The vascular system is sorted into an arteriolar and venular design. The 100 pictures are having a measurements of 1504 × 1000 comprise retinal courses, veins, AVR, and entire vascular structure for ground certainties. It likewise has an explanation at the picture level for HR.[32]

3.5.4 VICAVR

This dataset comprises of 58 retinal pictures. The dataset was utilized to register the supply route/vein proportion and the pictures are caught utilizing NW-100 Top Canon mydriatic camera with a focused optic plate and 768 × 584 pixels' goals. The database contains the subtleties of the vessel estimated from the optic plate at various radii alongside the sort of vessel (A/V proportion). The ground truth subtleties were watched by three picture examination specialists.[7]

3.5.5 INSPIRE AVR

INSPIRE AVR with 40 shading pictures of the vessels and optic circle and an arterial–venous proportion reference standard. The orientation standard is the normal of the appraisal of two specialists utilizing IVAN (a semi-mechanized PC program created by the University of Wisconsin, Madison, WI, USA) on the pictures.[20]

The retinal fundus image databases with the number of images available for HR classification are mentioned in Table 3.4.

3.6 PROPOSED METHOD

The retinal database of VICAVR (Fig. 3.12a–c) and STARE (Fig. 3.12d–f) is used in the method. The Figure 3.11 indicates the steps of the method proposed. The method takes the retinal images as input in the first step represented in Figure 3.12(i). Then the green channel of the fundus image is extracted. The next step is to enhance the retinal image using CLAHE. The next step is to localize the OD using morphological operations. Then it is to segment the BV and classifying them as arteries and veins. In the final step based on the ratio of AVR ratio, the HR classification is done.

TABLE 3.4 Retinal Database with Number of Fundus Images Available for HR.

SI. No	Database	Total images available
1	DRIVE	40
2	STARE	400
3	AVRDB	100
4	VICAVR	58
5	INSPIRE AVR	40

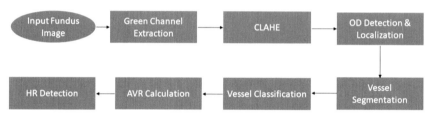

FIGURE 3.11 Proposed method.

3.6.1 GREEN CHANNEL

It is the second step of HR detection. The fundus image's green band is isolated because the contrast between the green channel BV's and the red and blue channels is more contrasted. The difference in intensity in the background is smaller in green plane of the fundus image. The green channel image is shown in Figure 3.12(ii).

3.6.2 CONTRAST LIMITED ADAPTIVE HISTOGRAM EQUALIZATION-CLAHE

CLAHE is often used in improving retinal image with low contrast. A transformation function per neighborhood pixel, derived from a minimal contrast procedure. CLAHE was introduced primarily to avoid the noise over amplification. The enhanced image is shown in Figure 3.12(iii).

3.6.3 OPTIC DISC LOCALIZATION

This section describes one of the major steps in diagnosing HR is to locate the OD. The nerves enter and leave the retina to the brain and travel through the OD from the brain to the retina. Thus the OD functions as an entry mark and a mark remains. The localization of OD is done by calculating the maximum intensity level of the average filtered fundus image and the region of interest is taken as four times the radius of the OD. The localized OD is shown in Figure 3.12(iv).

3.6.4 VESSEL SEGMENTATION AND CLASSIFICATION

OD localization helps in segmenting and separating the arteries and veins. Supervised and unsupervised techniques for classifying BV's are available. All of the supervised techniques implemented the pixel dependent classification. Neural networks and support vector machines are the main supervised strategies. The segmented BV's are shown in Figure 3.12(v). The approach mentioned uses the vessel classification neural network which is the supervised method. The adopted approach first trains the collection of STARE, VICAVR database training images and then checks the images to classify vessels as arteries or veins. The next step is to measure the arterial and vein widths. To measure the distance, take the counterpart of separate arteries and veins. By having the complement, it converts 0 pixel into 1 and 1 pixel into 0. Next calculate the distance by using the distance transform and morphological thinning method to calculate the approximate or accurate distance from each binary pixel of images to the nearest zero pixel. The gap would naturally be zero for null image pixels.

3.6.5 AVR CALCULATION AND HR DETECTION

The final step is to calculate the AVR ratio which is mentioned in the above sections. Based on the AVR ratio obtained the classification of HR is done which is mentioned in the Figure 3.3.

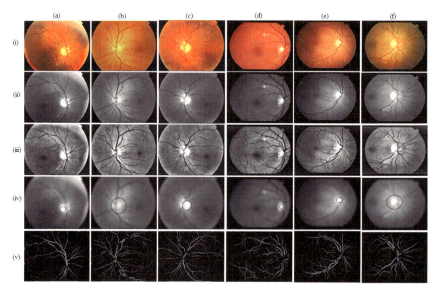

FIGURE 3.12 Output of the proposed method.

3.7 COMPARATIVE ANALYSIS

This section will give the Acc obtained for various algorithms for various databases. From the observation, it is found that Abbasi et al. used conventional approach for HR detection on locally available database and obtained a low Acc of 81%. Whereas Irshad et al. got very good results of 98.65% with the conventional methods using VICAVR database. Using ML approach for grading the HR Syahputra et al. achieved a highest Acc of 100% using a testing sample of 20 images from STARE database. Authors have used only one type of database for HR detection; but in the proposed method, VICAVR and STARE databases are used for HR detection. The Acc for HR grading using various algorithms is listed in Table 3.5 along with the database used.

3.8 CONCLUSION AND FUTURE WORK

Hypertensive eye ailment is distinguished by a robotized procedure from the retinal vein pictures. The veins are removed utilizing different techniques.

The talked about strategies pursue different strides from pre-handling to the division of veins and discovering AVR proportion. Each progression performed well for the improvement of the outcomes. The productivity of different calculations proposed is investigated by contrasting them and the techniques and datasets utilized.

TABLE 3.5 Comparison of Acc of Various Algorithms for HR Grading.

Author	Algorithms or performance metrics	Database	Acc (%)
Ortiz et al., 2010 [27]	Morphological operations	Hospital Universitario San	82
Manikis et al., 2011 [19]	AVR ratio	DRIVE	93.71
Ortíz et al., 2012 [28]	Morphological operations	Hospital Universitario	82
Irshad and Akram, 2014 [12]	AVR ratio	AVRDB	81.3
Khitran et al., 2014 [16]	Hybrid classifier	DRIVE	98
Abbasi and Akram, 2014 [1]	AVR ratio	Local data	81
Irshad et al., 2016 [13]	AVR ratio SVM	VICAVR	98.65
Syahputra et al., 2017 [33]	Probabilistic neural networks	STARE	100
Ahmad et al., 2018 [3]	AVR ratio	AVRDB	89.4
Savant and Shenvi, 2019 [30]	AVR ratio	DRIVE	86.67
Kiruthika et al., 2019 [17]	Radon vessel tracking algorithm	DRIVE	92.55
Triwijoyo et al., 2017 [34]	CNN	DRIVE	98.6
Akbar et al., 2018 [4]	AVR ratio	INSPIRE AVR	97.50
Akbar et al., 2018 [5]	AVR ratio	INSPIRE AVR	98.76
Proposed method	AVR ratio	VICAVR	98.5
		STARE	98.3

Many automatic techniques are accessible for HR discovery yet there is a requirement for such a framework which considers total fundus picture for programed HR location and evaluating. In this manner, this similar examination will be creative in automatic supported demonstrative arrangement of HR location and evaluating.

KEYWORDS

- retina
- blood vessels
- diabetic retinopathy
- hypertensive retinopathy
- machine learning

REFERENCES

1. Abbasi, U. G.; Akram, M. U. Classification of Blood Vessels as Arteries and Veins for Diagnosis of Hypertensive Retinopathy. In *2014 10th International Computer Engineering Conference (ICENCO)*; IEEE, 2014 Dec; pp 5–9.
2. Agurto, C.; Joshi, V.; Nemeth, S.; Soliz, P.; Barriga, S. Detection of Hypertensive Retinopathy Using Vessel Measurements and Textural Features. In *2014 36th Annual International Conference of the IEEE Engineering in Medicine and Biology Society*; IEEE, 2014 Aug; pp 5406–5409.
3. Ahmad, F.; Sial, M. R. K.; Yousaf, A.; Khan, F. Textural and Intensity Feature Based Retinal Vessels Classification for the Identification of Hypertensive Retinopathy. In *2018 IEEE 21st International Multi-Topic Conference (INMIC)*; IEEE, 2018 Nov; pp 1–4.
4. Akbar, S.; Akram, M. U.; Sharif, M.; Tariq, A.; Khan, S. A. Decision Support System for Detection of Hypertensive Retinopathy Using Arteriovenous Ratio. *Artif. Intell. Med.* **2018**, *90*, 15–24.
5. Akbar, S.; Akram, M. U.; Sharif, M.; Tariq, A.; Yasin, U. Arteriovenous Ratio and Papilledema Based Hybrid Decision Support System for Detection and Grading of Hypertensive Retinopathy. *Comput. Methods Programs Biomed.* **2018**, *154*, 123–141.
6. Ashley, E. A.; Niebauer, J. Chapter 6: Hypertension. In *Cardiology Explained*. London: Remedica, 2004.
7. Dashtbozorg, B.; Mendonça, A. M.; Campilho, A. An Automatic Graph-based Approach for Artery/Vein Classification in Retinal Images. *IEEE Trans. Image Process.* **2013**, *23* (3), 1073–1083.
8. Downie, L. E.; Hodgson, L. A.; DSylva, C.; McIntosh, R. L.; Rogers, S. L.; Connell, P.; Wong, T. Y. Hypertensive Retinopathy: Comparing the Keith–Wagener–Barker to a Simplified Classification. *J. Hyper.* **2013**, *31* (5), 960–965.
9. Erden, S.; Bicakci, E. Hypertensive Retinopathy: Incidence, Risk Factors, and Comorbidities. *Clin. Exp. Hyper.* **2012**, *34* (6), 397–401.
10. Grosso, A.; Veglio, F.; Porta, M.; Grignolo, F. M.; Wong, T. Y. Hypertensive Retinopathy Revisited: Some Answers, More Questions. *Br. J. Ophthalmol.* **2005**, *89* (12), 1646–1654.

11. Grossniklaus, H. E.; Geisert, E. E.; Nickerson, J. M. Introduction to the Retina. *Progress Mol. Biol. Trans. Sci.* **2015**, *134*, 383–396.
12. Irshad, S.; Akram, M. U. Classification of Retinal Vessels Into Arteries and Veins for Detection of Hypertensive Retinopathy. In *2014 Cairo International Biomedical Engineering Conference (CIBEC)*; IEEE, 2014 Dec; pp 133–136.
13. Irshad, S.; Ahmad, M.; Akram, M. U.; Malik, A. W.; Abbas, S. Classification of Vessels as Arteries Verses Veins Using Hybrid Features for Diagnosis of Hypertensive Retinopathy. In *2016 IEEE International Conference on Imaging Systems and Techniques (IST)*; IEEE, 2016 Oct; pp 472–475.
14. Irshad, S.; Salman, M.; Akram, M. U.; Yasin, U. Automated Detection of Cotton Wool Spots for the Diagnosis of Hypertensive Retinopathy. In *2014 Cairo International Biomedical Engineering Conference (CIBEC)*; IEEE, 2014 Dec; pp 121–124.
15. Senior, K. R. (Ed.). *The Eye: The Physiology of Human Perception*. The Rosen Publishing Group, Inc, 2010.
16. Khitran, S.; Akram, M. U.; Usman, A.; Yasin, U. Automated System for the Detection of Hypertensive Retinopathy. In *2014 4th International Conference on Image Processing Theory, Tools and Applications (IPTA)*; IEEE, 2014 Oct; pp 1–6.
17. Kiruthika, M.; Swapna, T. R.; Santhosh, K. C.; Peeyush, K. P. Artery and Vein Classification for Hypertensive Retinopathy. In *2019 3rd International Conference on Trends in Electronics and Informatics (ICOEI)*; IEEE, 2019 Apr; pp 244–248.
18. Latha, M. A.; Evangeline, N. C.; SankaraNarayanan, S. Colour Image Segmentation of Fundus Blood Vessels for the Detection of Hypertensive Retinopathy. In *2018 Fourth International Conference on Biosignals, Images and Instrumentation (ICBSII)*; IEEE, 2018 Mar; pp 206–212.
19. Manikis, G. C.; Sakkalis, V.; Zabulis, X.; Karamaounas, P.; Triantafyllou, A.; Douma, S.; Marias, K. An Image Analysis Framework for the Early Assessment of Hypertensive Retinopathy Signs. In *2011 E-Health and Bioengineering Conference (EHB)*; IEEE, 2011 Nov; pp 1–6.
20. Mendonça, A. M.; Sousa, A.; Mendonça, L.; Campilho, A. Automatic Localization of the Optic Disc by Combining Vascular and Intensity Information. *Comput. Med. Imag. Graph.* **2013**, *37* (5–6), 409–417.
21. Murugan, R. Implementation of Deep Learning Neural Network for Retinal Images. In *Handbook of Research on Applications and Implementations of Machine Learning Techniques*; IGI Global, 2020; pp 77–95.
22. Murugan, R. The Retinal Blood Vessel Segmentation Using Expected Maximization Algorithm. In *Computer Vision and Machine Intelligence in Medical Image Analysis*; Springer: Singapore, 2020; pp 55–64.
23. Narasimhan, K.; Neha, V. C.; Vijayarekha, K. Hypertensive Retinopathy Diagnosis from Fundus Images by Estimation of Avr. *Procedia Eng.* **2012**, *38*, 980–993.
24. Niemeijer, M.; Staal, J.; van Ginneken, B.; Loog, M.; Abramoff, M. D. Comparative Study of Retinal Vessel Segmentation Methods on a New Publicly Available Database. *Med. Imag. 2004: Image Process.* **2004**, *5370*, 648–656.
25. Nirmala, S. R.; Chetia, S. Retinal Blood Vessel Tortuosity Measurement for Analysis of Hypertensive Retinopathy. In *2017 International Conference on Innovations in Electronics, Signal Processing and Communication (IESC)*; IEEE, 2017 Apr; pp 45–50.

26. Ortíz, D.; Cubides, M.; Suárez, A.; Zequera, M.; Quiroga, J.; Gómez, J.; Arroyo, N. Support System for the Preventive Diagnosis of Hypertensive Retinopathy. In *2010 Annual International Conference of the IEEE Engineering in Medicine and Biology*; IEEE, 2010 Sept; pp 5649–5652.
27. Ortiz, D.; Cubides, M.; Suarez, A.; Zequera, M.; Quiroga, J.; Gómez, J. A.; Arroyo, N. System for Measuring the Arterious Venous Rate (AVR) for the Diagnosis of Hypertensive Retinopathy. In *2010 IEEE ANDESCON*; IEEE, 2010 Sept; pp 1–4.
28. Ortíz, D.; Cubides, M.; Suarez, A.; Zequera, M.; Quiroga, J.; Gómez, J. A.; Arroyo, N. System Development for Measuring the Arterious Venous Rate (AVR) for the Diagnosis of Hypertensive Retinopathy. In *2012 VI Andean Region International Conference*; IEEE, 2010 Sept; pp 53–56.
29. Prabhu, R. Understanding of Convolutional Neural Network (CNN)—Deep Learning, 2018. https://medium.com/@RaghavPrabhu/understanding-of-convolutional-neural-network-cnn-deep-learning-99760835f148
30. Savant, V.; Shenvi, N. Analysis of the Vessel Parameters for the Detection of Hypertensive Retinopathy. In *2019 3rd International conference on Electronics, Communication and Aerospace Technology (ICECA)*; IEEE, 2019 Jun; pp 838–841.
31. Akbar, S.; Hassan, T.; Akram, M. U.; Yasin, U.; Basit, I. AVRDB: Annotated Dataset for Vessel Segmentation and Calculation of Arteriovenous Ratio, 2017.
32. Akbar, S.; Akram, M. U.; Sharif, M.; Tariq, A.; Yasin, U. Decision Support System for Detection of Papilledema through Fundus Retinal Images. *J. Med. Syst.* **2017**, *41* (4), 66.
33. Syahputra, M. F.; Aulia, I.; Rahmat, R. F. Hypertensive Retinopathy Identification from Retinal Fundus Image Using Probabilistic Neural Network. In *2017 International Conference on Advanced Informatics, Concepts, Theory, and Applications (ICAICTA)*; IEEE, 2017 Aug; pp 1–6.
34. Triwijoyo, B. K.; Budiharto, W.; Abdurachman, E. The Classification of Hypertensive Retinopathy Using Convolutional Neural Network. *Procedia Comput. Sci.* **2017**, *116*, 166–173.
35. Carolina Ophthalmology, P. A. Diseases & Surgery of the Eye, Retina Center https://www.carolinaeyemd.com/retina-center-hendersonville/# [accessed 5 May 2020].
36. Walsh, J. B. Hypertensive Retinopathy: Description, Classification, and Prognosis. *Ophthalmology* **1982**, *89* (10), 1127–1131.
37. Wong, T. Y.; Mitchell, P. Hypertensive Retinopathy. *New Engl. J. Med.* **2004**, 351 (22), 2310–2317.
38. Zhang, B.; Zhang, L.; Zhang, L.; Karray, F. Retinal Vessel Extraction by Matched Filter with First-order Derivative of Gaussian. *Comput. Biol. Med.* **2010**, *40* (4), 438–445.
39. Rani, A.; Mittal, D. Measurement of Arterio-venous Ratio for Detection of Hypertensive Retinopathy through Digital Color Fundus Images. *J. Biomed. Eng. Med. Imag.* **2015**, *2* (5), 35–35.
40. Modi, P.; Arsiwalla, T. Hypertensive Retinopathy. In *StatPearls [Internet]*. StatPearls Publishing, 2019.
41. Hoover, A. STARE Database, 1975. http://www. ces. clemson. edu/~ ahoover/stare
42. Khare, N.; Devan, P.; Chowdhary, C. L.; Bhattacharya, S.; Singh, G.; Singh, S.; Yoon, B. SMO-DNN: Spider Monkey Optimization and Deep Neural Network Hybrid Classifier Model for Intrusion Detection. *Electronics* **2020**, *9* (4), 692.

43. Chowdhary, C. L.; Acharjya, D. P. Segmentation and Feature Extraction in Medical Imaging: A Systematic Review. *Procedia Comput. Sci.* **2020,** *167*, 26–36.
44. Reddy, T.; RM, S. P.; Parimala, M.; Chowdhary, C. L.; Hakak, S.; Khan, W. Z. A Deep Neural Networks Based Model for Uninterrupted Marine Environment Monitoring. *Comput. Commun.* **2020**.
45. Das, T. K.; Chowdhary, C. L.; Gao, X. Z. Chest X-Ray Investigation: A Convolutional Neural Network Approach. *J. Biomimetics, Biomater. Biomed. Eng.* **2020,** *45*, 57–70.

CHAPTER 4

Big Image Data Processing: Methods, Technologies, and Implementation Issues

U. S. N. RAJU*, SURESH KUMAR KANAPARTHI,
MAHESH KUMAR MORAMPUDI, SWETA PANIGRAHI, and
DEBANJAN PATHAK

*Department of Computer Science and Engineering,
National Institute of Technology Warangal, Telangana State, India*

*Corresponding author. E-mail: usnraju@nitw.ac.in

ABSTRACT

Big Image Data Processing (BIDP) refers to the processing of images that are huge in terms of quantity, individual dimension, and individual size with respect to memory. This chapter elaborates on methods to deal with the three above-mentioned categories of images. In these scenarios, the data can be stored using a Distributed File System. To work with this amount of data, different programing paradigms can be used such as Hadoop's MapReduce, Matlab's MapReduce, and "Hadoop-Matlab" integrated environment with MapReduce Programing. The authors formed a Hadoop cluster with 116 systems and processed 1.2 TB of text data for word count task. The authors have also performed image retrieval on Corel 1000, Corel 10,000, Brodatz Textures, Mirflickr and ImageNet datasets effectively with this cluster configuration. The authors have created and processed a 32768 × 32768 dimension image and a 3.14 GB image using the MapReduce paradigm. Different applications using these technologies and methods are image retrieval and object detection, which can be used in a multiresolution environment as well.

4.1 INTRODUCTION

4.1.1 IMAGE AND VIDEO PROCESSING AND APPLICATIONS

An image is made up of picture elements, called pixels or image elements. Each numeric value of pixel represents its intensity and can be represented by spatial coordinates with *x* and *y* on the *x-axis* and *y-axis*, respectively. An image can be in binary, grayscale, or color format. Digital image processing implies processing with an image on a digital computer.[1] Digital video is acquired by a time sequence of two-dimensional spatial intensity arrays.[2] In simple words, images or frames are displayed at a rate of 72 frames per second to make it look continuous. Image and video processing has their application in many fields and its demand is growing with the thriving technology. Some of its applications are stated in the proceeding section.

One of the applications of image processing is in restoration.[3] It is used for modification of images to remove noise and improve the image quality, this, in turn, is advantageous for detection and retrieval applications. The medical field has numerous applications wherein image processing is used such as X-rays[4] and CT scans.[5] Image and video processing are used in object detection.

Object detection not only classifies, but also gives a precise location of the object in each image or frame of a video. At the same time, object detection is a fundamental problem of computer vision that has applications in image classification,[6] human behavior analysis[7], and autonomous driving.[8] In computer vision, many robotic machines perform image-based tasks.[9] Robots employ image processing methods to track ways such as line follower robot and detection of a hurdle. Another application is pattern recognition where it is combined with artificial intelligence for recognition,[10] modeling and segmentation.[11] In the field of video processing, video surveillance contributes hugely to Big Video data. Monitoring surveillance videos is very crucial for protection and security in many metropolitan cities.[12] Big Video processing requires efficient video compression and transmission to enable smart cities with Internet of Things technique which further helps in monitoring human activity information.[13] Video tracking for suspicious vehicles, movements, etc., is an application of video processing in a large scale.[14] Another area where Big Video is applied is in transportation management. The development of an intelligent transportation system requires processing in a Big Data environment.[15] Managing the transport system gives rise to issues like monitoring vehicle density and traffic congestion wherein a large quantity of videos needs to be processed using Big Data mechanisms.[16]

4.1.2 BIG DATA

Big Data is a means of describing data problems that cannot be solved by traditional tools. For better comprehension of Big Data problems, initially in the early 2000s, it is accepted that Big Data can be characterized by three Vs, that is, Volume, Variety, and Velocity. But Big Data goes beyond these three Vs. To prepare for the advantages and challenges of Big Data initiatives, it is characterized by seven more Vs.[17–19] All these ten Vs are explained in this section.

Volume: Volume is the quantity of data that we have. With an increase in the number of new technologies and devices, there is an exponential growth in data. These data can be extremely valuable if it can be utilized in a proper manner. About 90% of all data ever created was generated in the past 2 years. Therefore, the scale is what makes Big Data big.

Variety: One of the biggest challenges faced by Big Data is Variety. Most of the data generated are unstructured which includes various types of data from XML data to tweets, photos, and videos. Organization of this data in a semantic way is difficult as the data itself is rapidly changing.

Velocity: Velocity denotes the speed at which data creation increases and the speed at which relational databases can store, process, and analyze data. The promises of real-time data processing attract interest as it allows companies to achieve tasks such as displaying personalized advertisements on the websites visited in accordance with a person's recent history of search, viewing, and purchase.

Veracity: Veracity states to make the data accurate. The value of Big Data ceases if it is not accurate, which requires discarding the noise before beginning analysis. The simplest example is the contacts that enter your marketing automation system with false names and inaccurate contact information.

Value: Big Data has a huge potential value. Even though, discarding poor data's cost is also huge. Because data are actually worthless unless it is analyzed to get accurate data and information provided by it.

Visualization: Once the data have been processed, it needs to be presented in an accessible and readable manner. Visualization can contain loads of parameters and variables. This has become one of the challenges of Big Data.

Variability: Variability differs from variety. A restaurant may have 20 different kinds of food items on the menu. However, if the same item from the menu tastes different each day, then it is called variability. The same applies to data, whenever the meaning of a data changes constantly it affects the homogeneous nature of data. Variability indicates data whose meaning constantly changes.

Vulnerability: With a huge amount of data, there also arise concerns about security. A data breach on Big Data can cause an exploitation of important information. Many hackers have attempted and succeeded in many Big Data breaches.

Volatility: Before the advent of Big Data, data were stored indefinitely. But due to the volume and velocity of Big Data, volatility needs to be considered. It needs to be established that how long data should be stored and when to consider that data have become irrelevant or historic.

Validity: Validity refers to how accurate and correct the data is for its intended use. Benefits from Big Data can be derived if the underlying data are consistent in quality, metadata, and common definitions.

4.1.3 BIG IMAGE DATA PROCESSING

The demand for processing an enormous number of images, images of large dimension and images big in size made the authors explore the new technologies, which can accomplish this.[20,21] In this process, "Big Image/Video Data Processing" has evolved as shown in Figure 4.1. The relation between Big Data and Image Processing is shown in Figure 4.2. Big Image/Video data processing has solved many technological challenges which including storage, compression, analysis, transmission, and recognition.[22–25] Big Image/Video Data processing plays an important role in fulfilling modern-day technical demands such as intelligent transport system,[15] big image classification and retrieval,[26] human behavior monitoring.[27]

4.1.4 CATEGORIES OF BIG IMAGE DATA PROCESSING

The general perception of Big Image Data Processing is that it deals with the processing of images that are huge in quantity. However, Big Image implies

(1) images which are large in quantity, shown in Figure 4.3. (2) individual image big with respect to Dimension (M × N) as shown in Table 4.1 and (3) individual images big with respect to Size, that is, amount of storage required to store it, as shown in Table 4.2.

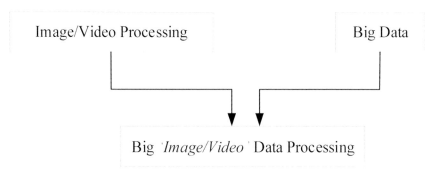

FIGURE 4.1 Evolving of "Big Image Data Processing."

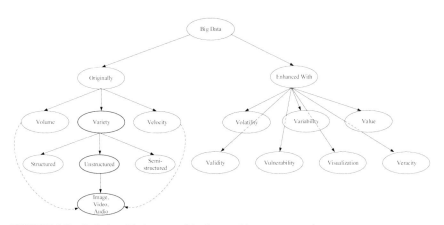

FIGURE 4.2 Relationship between big data and image processing.

In this chapter, the authors have given details about how to handle these three types of "big image" to store and process in a distributed environment.

In this chapter, the authors have given different methods, technologies and implementation issues that they have experienced in making BIDP success.

- The objectives of this chapter are:
- To give different methods of handling Big Image Data.

- To discuss different technologies that can be used for processing the Big Image Data.
- To share the experience with respect to the implementation.
- To give the applications of Big Image Data Processing.

FIGURE 4.3 A large number of images.

4.2 BACKGROUND

The number of images or videos to be processed is not just huge in quantity but also has enormous size and dimension. Therefore, given the existing technologies and environment, it is not possible to process this data without compromising on time. In the following section, the authors have presented a scenario leading to motivation for Big Image/Video data and methods to process it.

TABLE 4.1 Examples of Dimension Based Huge Images.

Image Name	Image	Dimension
Galaxy Image		1.5 billion pixel
NewYork city		203200×101600≈20 gig pixels
Sky		100000×50000 pixels
Tokyo Tower		45 gig pixels
Roppongi Hills Mori Tower.		150 gig pixels

TABLE 4.2 Examples of Size Based Huge Images.

Image Name	Image	Size
Galaxy Image		4.3 GB
The Garden of Earthly Delights by Bosch		5.7 GB
Louise Elisabeth Vigee-Lebrun-Marie-Antoinette-Google Art Project		2.7 GB
Hans Holbein the Younger - The Ambassadors - Google Art Project		3.0 GB

Video Assignments: Today, students are able to use electronic devices without any difficulty. Therefore, the submission of assignments in video format can be done comfortably by them. Raju et al.[28] proposed a new technique to review the assignments submitted by the students. They are required to record a video of themselves explaining the given problem.[29] This process helps the students to improve their learning capabilities. This can also prove useful for science students pursuing under graduation where they are required to learn theorems and construe proofs. Another benefit is that it facilitates teachers to raise discussion topics that allow students to work out their topics. Some of the video assignments submitted by students are shown in Figure 4.4. The problem that the authors have faced is some of the students, by using video cutting software, copied it from their friends.

To handle a large number (~300) of videos to find video plagiarism which is a tedious and time-consuming process.[30] So the authors have thought of processing these videos with the help of state-of-the-art technology: Hadoop with MapReduce paradigm.

IEEE BigMM Conference: The conference IEEE BigMM[31] which has started in 2015 is also another motivating factor by the authors. BigMM stands for Big Multimedia. The target of the conference is to invite papers, which are in the domain of Multimedia data satisfying the characteristics of Big Data.

Surveillance Videos: Applications of Big Data consume a lot of space in the research area and industry. Video streams coming from CCTV cameras is one of the main contributing and important cause among other sources of Big Data. Surveillance videos highly contribute to unstructured Big Data. CCTV cameras are installed in many places having demands for security. Surveillance ability and improved security are not possible without technology. Many technical innovations have come into existence, such as access control devices, video surveillance, and alarms. In a survey, it was found that all the respondents have either a system of video surveillance installed which is 95% or are planning to install the system in the next 1 year which is 5%. One respondent has reported the largest number of cameras, totaling up to 25000. Certainly, in each network, there has been an increase of almost 70% of the average number of cameras. The year from 2015 to 2018 saw an increase from about 2900 to 4900 cameras. The newest survey suggests reports that 20% of respondents have 10000 or more cameras, whereas just 5% of them had in the previous survey.[32]

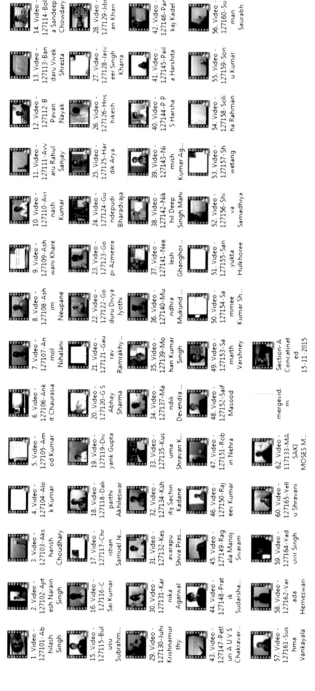

FIGURE 4.4 Assignments submitted by the students in the form of video.

4.2.1 EXISTING TECHNOLOGIES:

The main concerns while dealing with processing of Big Image Data are (1) storage of the given data when it cannot be stored in the existing infrastructure and (2) processing of the given data when it cannot be processed with the existing infrastructure. In some cases, both can be done with the existing infrastructure, but it is a very time-consuming process. So to deal with this, the authors have discussed different existing technologies in this section.

A. *Hadoop:* Hadoop is a part of the Apache project.[33] It is an open-source Java-based framework used for storage and processing of Big Data in a distributed environment.
   - **Storage:** Hadoop mainly contains two parts. One for storage and another for processing. For storing the data, it uses a file system known as Hadoop Distributed File System (HDFS). If the amount of data cannot fit into the memory of a single computer, a Hadoop cluster can be made with n number of computers, which gives combined storage. The total storage that can be contributed by all the computers in the cluster is termed as HDFS. In this scenario, all the computers which are a part of the cluster can access the data.
   - **Processing:** As the data are stored in a distributed file system, a different programing paradigm is needed to process these data. So, Hadoop uses the MapReduce programing paradigm for it. When dealing with large data, the MapReduce paradigm is one of the best solutions to get the results in less time than that on doing it on a single system. This is a programing paradigm in which the execution takes place where the data reside. The execution takes place in three stages: Map, Shuffle & Sort and Reduce stages. The Map stage takes in the input in *<Key, Value>* pair and produces the output also as *<Key, Value>* pair. Then the Shuffle & Sort stage will sort this based on the "*key*". Therefore, the reducer will consolidate the work for each of the *key* and produce the final output. For storing the data in intermediate steps Distributed File System can be used. This data can be in any form: Text, Images, Videos, Log Data, etc.

B. *MATLAB with Matlab Distributed Computing Server (MDCS)*: The MATLAB Distributed Computing Server (MDCS) allows users to submit (from within MATLAB) sequential or parallel MATLAB code to a cluster. Before working on a distributed environment, MDCS should be installed on all the systems of the cluster. The authors have used a 96 workers (or cores) MDCS setup in their lab where it is installed on 24 system cluster where each system is of 4 cores. The parallel computing toolbox which contains "parpool" is installed on the head node which is one among the 24 system cluster. So, one of the 24 systems is considered as both head node and client node. "parpool" can be used for the execution in different cores of a system. The "parpool" can be used in two different modes: (1) on local mode, a single system with available cores on it (2) by using the cores from all the systems in the cluster of systems (*gcp*). The integration of Matlab with Hadoop is done so that the data can be read from HDFS by parpool.[34]

C. *Spark*: Apache Spark is a distributed cluster-computing general-purpose open-source framework. Spark offers an interface for execution in the whole cluster facilitating fault tolerance and data parallelism. Spark was developed at the University of California, Berkeley's AMPLab. Later, the Spark codebase was given to the Apache Software Foundation. Resilient Distributed Dataset (RDD) which is a read-only distribution of multiset data items over a cluster of computers is the architectural foundation of Spark.[35] When Hadoop's MapReduce is used with multiple jobs to complete the given task, the intermediate results after completion of every job are going to be stored in HDFS. Reading the data from HDFS for the next job is a time-consuming process. However, in Spark, the intermediate results are stored in memory, so reading the data for the next job saves a lot of time. This, in turn, becomes faster by several orders of magnitude compared to Apache Hadoop MapReduce implementation.[35-37] As Spark provides a faster environment, implementation of iterative algorithms, which accesses their dataset in a loop for multiple times and interactive data analysis, which has repeated querying of data in the database is facilitated. Among the group of iterative algorithms are the training algorithms of the machine learning systems, which gave the initial motivation for the development of Apache Spark.[38]

To process Big Image Data, the above-given technologies can be used individually or in combination. Sarmad Istephan et al.[39] proposed a method to retrieve an image from unstructured medical image Big Data with a case study on epilepsy. They have used two types of criteria to validate the feasibility of the proposed framework: accuracy and ability. The accuracy is tested by executing the query on data that contains both structured and unstructured data. To test the ability of the framework, the results are compared by executing the query on different sized Hadoop clusters. The same kind of ability is tested in[40] also. One novel CBIR framework was proposed by Lan Zhang et al.,[41] known as PIC, where cloud computing is used for searching an image from a large image dataset while securing the privacy of input data. Here to deal with massive images, they have designed a system suitable for distributed and parallel computation to expedite the search process. Le Dong[42] proposed an effective processing framework named Image Cloud Processing (ICP) to deal with data explosion in the image processing field. The ICP framework consists of two mechanisms: Static ICP (SICP) and Dynamic ICP (DICP), where SICP is designed to cooperate with MapReduce paradigm and DICP implemented through a parallel processing procedure works with the traditional processing mechanism of the distributed system. To validate the ICP framework, they have used the ImageNet dataset. Jiachen Yang[43–47] have used maximal mutual information criterion to reduce the feature vector dimension to decrease the retrieval time.

4.3 MAIN FOCUS OF THE CHAPTER

The objectives of this section are to discuss different methods to handle any of these categories with respect to Big Images.

- A large number of Images (*SequenceFile*).
- A single large image with higher dimensions (*Make into small pieces with respect to dimension and do MapReduce on them*).
- A single large image with huge memory (*Make into small pieces with respect to memory and do MapReduce on them*).

4.3.1 METHODS FOR PROCESSING BIG IMAGE DATA

To achieve the above-mentioned objectives with respect to Big Image Data (BID), the MapReduce programing paradigm is used in three different ways: *local*, *Hadoop cluster*, and Matlab's *Parallel Pool.* To work with this, the different options that can be used are: *0*, *cluster*, and *gcp* (get current parallel pool), respectively, in the implementation code as cluster setup. When "0" is used, the data will be taken from the local system and Matlab's MapReduce will do the entire job. When "*cluster*" is used as the option, the Hadoop's MapReduce will be active and data can be accessed from HDFS. Last, if "*gcp*" is the option used, Matlab's Parallel pool with MapReduce will be activated and the data can be taken from HDFS.

4.3.2 TECHNOLOGIES AND IMPLEMENTATION ISSUES FOR PROCESSING BID

It is necessary to deal with a large number of small files in BIDP, which is one of the main drawbacks of Hadoop and other distributed processing technologies. Processing large no. of small files creates large no. of memory references and that generates a lot of overhead for name node in Hadoop. Besides, more number of mappers is needed for more number of files. Sequence file format solves the problem of processing too many small files. Many small files are clubbed into a single sequence file which is used for processing as input for MapReduce programs.

The concept of Sequence File is putting each small file into a larger single file. Sequence files are binary files containing key-value pairs. They can be compressed at the record (key-value pair) or block levels. Because sequence files are binary, they have faster read/write than text-formatted files. Beyond packaging files into a manageable size, sequence files support compression of the keys, the values or both. So the type of compression determines the sequence file format.

- Uncompressed (neither Key nor the Value is compressed, that is, Key/value records are uncompressed)
- Record compressed (only values are compressed, key is not compressed)
- Block compressed (Both keys & values are compressed)

In all the implementations given here, the first MR job is to convert the given image files into sequence files. It can be done with Hadoop(*.seq files*) environment and as well as Matlab(*.mat file*). Before looking into the process of handling images, the process of handling big text data is discussed.

A. *Working with Text Data*: In this method, the authors created a total of 116 nodes Hadoop cluster. One of the nodes is considered as the master node whereas the remaining 115 nodes are considered as the slave nodes. These 116 nodes are situated in three different labs of their college. Table 4.3 shows the configuration of all the nodes in this cluster. The configuration capacity has become 28.13 TB with the help of this cluster. The authors have uploaded text data of a total size 1.2 TB with a replication factor of 5 into HDFS to test the cluster performance. Then execution of the standard "word-count" example is carried out on this data. The time taken for completion is 8 min 8 sec. The authors have also uploaded the entire process into youtube given in the link: https://www.youtube.com/watch?v=CSryEIkNGdk. The sample code for this one is given here.

TABLE 4.3 Hadoop Cluster Configuration with 116 Systems.

Node type	Ram size	Processor	CPU cores	Processor speed	Operating system	Hadoop version	Location
Master	8GB	Intel i7-4790	8	3.60GHz	Ubuntu (14.04)-64	2.7.2	Lab1
Slave1- Slave30	8GB	Intel i7-4790	8	3.60GHz	Ubuntu (14.04)-64	2.7.2	Lab1
Slave31- Slave68	4GB	Intel i7-4770	8	3.40GHz	Ubuntu (14.04)-64	2.7.2	Lab2
Slave69- Slave115	4GB	Intel i7-4770	8	3.40GHz	Ubuntu (14.04)-64	2.7.2	Lab3

```
%word_count.m
mapreducer(0);
datafolder = '/input';
files = fullfile(datafolder, '*.txt');
ds = datastore(files,'TextscanFormats' , '%s', 'Delimiter', ' ',
    'ReadVariableNames', false, 'VariableNames', 'Word');
output_folder = '/output';
outds = mapreduce(ds, @mapCountWords, @reduceCountWords, 'Output
    Folder', output_folder); readall(outds)
```

```
%mapCountWords.m
function mapCountWords(data, info, intermKVStore)
x = table2array(data);
for i=1:size(x,1)
    disp([string(x(i,1)) 1]); % displaying the key value pair
          %which is output of mapper
    add(intermKVStore,string(x(i,1)),1);
end
end
```

```
%reduceCountWords.m
function reduceCountWords(intermkey, intermValIter, outKVStore)
sum_occurences = 0;
while(hasnext(intermValIter))
    sum_occurences = sum_occurences + getnext(intermValIter)
end
add(outKVStore, intermkey, sum_occurences);
end
```

B. *Working with Large Number of Images*: The authors have integrated MATLAB with Hadoop and then built a cluster with 1-Master and 110-Slave Nodes (5-nodes of the original cluster were removed due to memory limitations during MATLAB installation). The authors worked on the CBIR problem by considering different standard image datasets: Corel 1000, Corel 10000, Brodatz Textures, Mirflickr (1,000,000 images), and ImageNet (1,281,167 images). Three MR Jobs are used in this process and they are given in Figures 4.5–4.7. The entire process is given in Algorithm-1.

Big Image Data Processing: Methods, Technologies 85

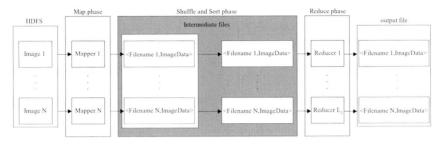

FIGURE 4.5 Outline of MapReduce job 1.

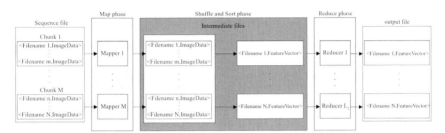

FIGURE 4.6 Outline of MapReduce job 2.

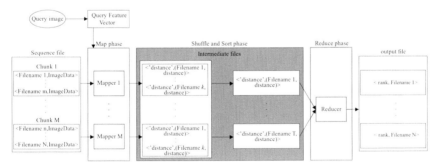

FIGURE 4.7 Outline of MapReduce job 3.

Algorithm-1:

Begin

    Step-1: Store all the images of the dataset into HDFS.

    Step-2: Give all the images to MR_Job1, which gives the <FileName, ImageData> as the output of this Job in the form of sequence file.

Step-3: The resultant sequence file of Step-2 is given as input to MR_Job2, which results <FileName, FeatureVector> as the output.

Step-4: For MR-Job3, the result of Step-3 is given as input along with the query Image which results in <Rank, Filename> with respect to the query image.

Step-5: Calculate the performance measures from the rank matrix resulted in Step-4.

End

C. *Working with Big Image of Huge Dimension:* The authors have created an image in Adobe Photoshop of dimension 32768 × 32768 shown in Figure 4.8. In their lab, with an 8 GB RAM i7 processor system, it took around 1 hr 21 min to process, that is, *to count the number of rectangles* in it. So, the authors used Hadoop with MATLAB for processing this by MapReduce model. The authors divided the 32768 × 32768 image into 1024 pieces of size 1024 × 1024 as shown in Figure 4.9(a), and 512 pieces are of size 2048 × 2048 as shown in Figure 4.9(b). This was stored into HDFS and MapReduce model was applied to count the number of rectangles. It took around 12 min to create a sequence file and 1 min 15 sec to complete the process of counting the number of rectangles.

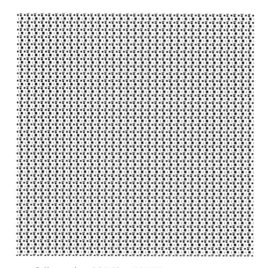

FIGURE 4.8 Image of dimension 32768 × 32768.

 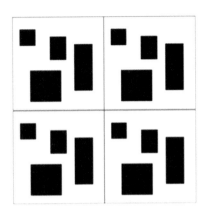

FIGURE 4.9 (a) 1024 sized block (b) 2048 sized block of Figure 4.8.

D. *Working with Big Image of Huge Size:* An image of a size of 3.14 GB was created by the authors. The authors tried to upload the image into an image processing software, but it was not readable and it showed an error *OUT OF MEMORY*. Therefore, the authors have divided the image into different pieces. As the author's HDFS chunk size is 64 MB, each piece was of size less than 64 MB. So, the image was divided into 100 blocks. Time elapsed for Job-1 completion is 3 min 30 sec and Job-2 completion is only 50 sec.

4.4   FUTURE RESEARCH DIRECTIONS

MapReduce or RDD can be used for distributed computing, similarly, the same MapReduce or RDD can be done on Parallel GPUs instead of a cluster of computers with only CPU. This will be even faster in completing the executions in applications like CBIR, CBIR for Multiresolution image datasets and Object Detection.

4.5   CONCLUSION

The authors have shown the process of handling Big Image Data. Three different cases are shown: (1) to handle a large number of images (2) Working with Big image of huge Dimension, and (3) Working with Big image of huge Size. The authors have processed different standard image datasets which are large in quantity to achieve image retrieval tasks using MapReduce paradigm by storing the data in a distributed file system. The different modes of parallel execution are discussed. The advantage of converting the files into sequence files is also discussed.

KEYWORDS

- **big image data**
- **distributed file system**
- **Hadoop**
- **Spark**
- **Matlab**
- **gigantic**

REFERENCES

1. Gonzalez, R. C.; Woods, E. W. *Digital Image Processing*, 4th ed.; Pearson: New York, 2018.
2. Bovik, A. C. *Handbook of Image and Video Processing*; Academic Press, 2010.
3. Papyan, V.; Elad, M. Multi-scale Patch-based Image Restoration. *IEEE Trans. Image Proces.* **2015,** *25* (1), 249–261.
4. Manzke, R.; Meyer, C.; Ecabert, O.; Peters, J.; Noordhoek, N. J.; Thiagalingam, A.; Reddy, V. Y.; Chan, R. C.; Weese, J. Automatic Segmentation of Rotational X-ray Images for Anatomic Intra-procedural Surface Generation in Atrial Fibrillation Ablation Procedures. *IEEE Trans. Med. Imag.* **2009,** *29* (2), 260–272.
5. Yang, W.; Zhong, L.; Chen, Y.; Lin, L.; Lu, Z.; Liu, S.; Wu, Y.; Feng, Q.; Chen, W. Predicting CT Image from MRI Data through Feature Matching with Learned Nonlinear Local Descriptors. *IEEE Trans. Med. Imag.* **2018,** *37* (4), 977–987.
6. Ma, X.; Schonfeld, D.; Khokhar, A. A General Two-dimensional Hidden Markov Model and Its Application in Image Classification. *2007 IEEE Int. Conf. Image Process.* **2007,** *6*, VI–41.
7. Cao, Z.; Simon, T.; Wei, S. E.; Sheikh, Y. Realtime Multi-person 2D Pose Estimation Using Part Affinity Fields. In *Proceedings of the IEEE Conference on Computer Vision and Pattern Recognition*; 2017; pp 7291–7299.
8. Chen, X.; Ma, H.; Wan, J.; Li, B.; Xia, T. Multi-view 3D Object Detection Network for Autonomous Driving. In *Proceedings of the IEEE Conference on Computer Vision and Pattern Recognition*; 2017; pp 1907–1915.

9. Ayache, N. Epidaure: A Research Project in Medical Image Analysis, Simulation, and Robotics at INRIA. *IEEE Trans. Med. Imag.* **2003,** *22* (10), 1185–1201.
10. Tao, D.; Li, X.; Wu, X.; Maybank, S. J. General Tensor Discriminant Analysis and Gabor Features for Gait Recognition. *IEEE Trans. Pattern Analy. Mach. Intell.* **2007,** *29* (10), 1700–1715.
11. Todorovic, S.; Ahuja, N. Unsupervised Category Modeling, Recognition, and Segmentation in Images. *IEEE Trans. Pattern Analy. Mach. Intell.* **2008,** *30* (12), 2158–2174.
12. Shao, Z.; Cai, J.; Wang, Z. Smart Monitoring Cameras Driven Intelligent Processing to Big Surveillance Video Data. *IEEE Trans. Big Data* **2017,** *4* (1), 105–116.
13. Tian, L.; Wang, H.; Zhou, Y.; Peng, C. Video Big Data in Smart City: Background Construction and Optimization for Surveillance Video Processing. *Future Gen. Comput. Syst.* **2018,** *86,* 1371–1382.
14. Subudhi, B. N.; Rout, D. K.; Ghosh, A. Big Data Analytics for Video Surveillance. *Multimedia Tools App.* **2019,** *78* (18), 26129–26162.
15. Hao, Q.; Qin, L. The Design of Intelligent Transportation Video Processing System in Big Data Environment. *IEEE Access* **2020,** *8,* 13769–13780.
16. Kadaieaswaran, M.; Arunprasath, V.; Karthika, M. Big Data Solution for Improving Traffic Management System with Video Processing. *Int. J. Eng. Sci.* **2017,** *7* (2).
17. The 7 V's of Big Data. Retrieved from https://impact.com/marketing-intelligence/7-vs-big-data/
18. Understanding Big Data: The Seven V's. https://dataconomy.com/2014/05/seven-vs-big-data/
19. The 10 Vs of Big Data. https://tdwi.org/articles/2017/02/08/10-vs-of-big-data.aspx/
20. Editorial Service, D. T. *Big Data Black Book*; Dreamtech Press: New Delhi, 2016.
21. Chuck, L. *Hadoop in Action*; Dreamtech Press: New Delhi, 2015.
22. Wang, W.; Zhao, W.; Cai, C.; Huang, J.; Xu, X.; Li, L. An Efficient Image Aesthetic Analysis System Using Hadoop. *Sign. Process.: Image Commun.* **2015,** *39,* 499–508.
23. Lin, Y.; Lv, F.; Zhu, S.; Yang, M.; Cour, T.; Yu, K.; Huang, T. Large-scale Image Classification: Fast Feature Extraction and SVM Training. *In CVPR 2011*; IEEE, 2011 June; pp 1689–1696.
24. Shiliang, Z.; Ming, Y.; Xiaoyu, W.; Yuanqing, L.; Qi, T. Semantic-aware Co-indexing for Image Retrieval. *IEEE Trans. Pattern Analy. Mach. Intell.* **2015,** *37* (12), 1673–1680.
25. Le, D.; Zhiyu, L.; Yan, L.; Ling, H.; Ning, Z.; Qi, C.; Xiaochun, C.; Ebroul, I. A Hierarchical Distributed Processing Framework for Big Image Data. *IEEE Trans. Big Data* **2016,** *2* (4), 297–309.
26. Tong, X. Y.; Guo, C.; Cheng, H. Multi-source Remote Sensing Image Big Data Classification System Design in Cloud Computing Environment. *Int. J. Internet Manuf. Serv.* **2020,** *7* (1–2), 130–145.
27. Yamamoto, N. Judging Students' Learning Style from Big Video Data Using Neural Network. In *International Conference on Emerging Internetworking, Data & Web Technologies*; Springer: Cham, 2020 Feb; pp 1–6.
28. Raju, U. S. N.; Kadambari, K. V.; Reddy, P. V. S. A New Method of Assessing the Students Using Video Assignments. In *2015 IEEE Global Engineering Education Conference (EDUCON)*; IEEE, 2015 Mar; pp 771–773.

29. Regina, A.; Peter, D.; Louise, K.; Andrew, L.; Jacqui, S. Thinking Creatively about Video Assignment- A Conversation with Penn Faculty. http://wic.library.upenn.edu/wicideas/facvideo.html
30. Raju, U. S. N.; Chaitanya, B.; Kumar, K. P.; Krishna, P. N.; Mishra, P. Video Copy Detection in Distributed Environment. In *2016 IEEE Second International Conference on Multimedia Big Data (BigMM)*; IEEE, 2016 Apr; pp 432–435.
31. BigMM 2020. http://bigmm2020.org/
32. International Trends in Video Surveillance- Public Transport Gets Smarter, 2018. https://www.uitp.org/sites/default/files/cck-focus-papers-files/1809-Statistics%20Brief%20-%20Videosurveillance-Final.pdf
33. Welcome to Apache™ Hadoop®!. http://hadoop.apache.org/
34. Getting Started with MapReduce. https://in.mathworks.com/help/matlab/import_export/getting-started-with-mapreduce.html
35. Zaharia, M.; Chowdhury, M.; Franklin, M. J.; Shenker, S.; Stoica, I. Spark: Cluster Computing with Working Sets (PDF). In *USENIX Workshop on Hot Topics in Cloud Computing* (HotCloud); 2014.
36. Zaharia, M.; Chowdhury, M.; Das, T.; Dave, A.; Ma, J.; McCauly, M.; … Stoica, I. Resilient Distributed Datasets: A Fault-tolerant Abstraction for In-memory Cluster Computing. In *Presented as Part of the 9th {USENIX} Symposium on Networked Systems Design and Implementation ({NSDI} 12)*; 2012; pp 15–28.
37. Xin, R. S.; Rosen, J.; Zaharia, M.; Franklin, M. J.; Shenker, S.; Stoica, I. Shark: SQL and Rich Analytics at Scale. In *Proceedings of the 2013 ACM SIGMOD International Conference on Management of Data*; 2013, Jun; pp 13–24.
38. Harris, D. 4 Reasons Why Spark Could Jolt Hadoop Into Hyperdrive. *Gigaom*, 2014. https://gigaom.com/2014/06/28/4-reasons-why-spark-could-jolt-hadoop-intohyperdrive
39. Sarmad, I.; Mohammad-Reza, S. Unstructured Medical Image Query Using Big Data- An Epilepsy Case Study. *J. Biomed. Info.* **2016**, *59*, 218–226.
40. Raju, U. S. N.; Suresh Kumar, K.; Haran, P.; Boppana, R. S.; Kumar, N. Content-based Image Retrieval Using Local Texture Features in Distributed Environment. *Int. J. Wavelets, Multiresol. Info. Process.* **2019**, 1941001.
41. Lan, Z.; Taeho, J.; Kebin, L.; Xiang-Yang, L.; Xuan, D.; Jiaxi, G.; Yunhao, Liu. PIC: Enable Large-scale Privacy Preserving Content-based Image Search on Cloud. *IEEE Trans. Parallel Dist. Syst.* **2017**, *25* (11), 3258–3271.
42. Le, D.; Zhiyu, L.; Yan, L.; Ling, H.; Ning, Z.; Qi, C.; Xiaochun, C.; Ebroul, I. A Hierarchical Distributed Processing Framework for Big Image Data. *IEEE Trans. Big Data* **2016**, *2* (4), 297–309.
43. Das, T. K.; Chowdhary, C. L.; Gao, X. Z. Chest X-Ray Investigation: A Convolutional Neural Network Approach. *J. Biomimetics, Biomater. Biomed. Eng.* **2020**, *45*, 57–70).
44. Chowdhary, C. L.; Acharjya, D. P. Segmentation and Feature Extraction in Medical Imaging: A Systematic Review. *Procedia Comput. Sci.* **2020**, *167*, 26–36.
45. Khare, N.; Devan, P.; Chowdhary, C. L.; Bhattacharya, S.; Singh, G.; Singh, S.; Yoon, B. SMO-DNN: Spider Monkey Optimization and Deep Neural Network Hybrid Classifier Model for Intrusion Detection. *Electronics* **2020**, *9* (4), 692.

46. Reddy, T.; RM, S. P.; Parimala, M.; Chowdhary, C. L.; Hakak, S.; Khan, W. Z. A Deep Neural Networks Based Model for Uninterrupted Marine Environment Monitoring. *Comput. Commun.* **2020**.
47. Jiachen, Y.; Bin, J.; Baihua, L.; Kun, T.; Zhihan, L. A Fast Image Retrieval Method Designed for Network Big Data. *IEEE Trans. Ind. Info.* **2017,** *13* (5), 2350–2359.

CHAPTER 5

N-grams for Image Classification and Retrieval

PRADNYA S. KULKARNI[1,2]

[1]*School of Computer Engineering and Technology, MIT World Peace University, Pune, India*

[2]*Honorary Research Fellow, Federation University, Australia*

*Corresponding author. E-mail: pradnya.kulkarni@mitwpu.edu.in

ABSTRACT

Content-based image retrieval (CBIR) and classification algorithms require features to be extracted from images. Global and low level image features such as color, texture, and shape fail to describe pattern variations within regions of an image. Bag of Visual Words approaches have emerged in recent years that extract features based on local pattern variations. These approaches typically outperform global feature methods in classification tasks. Recent studies have shown that Word N-Gram models common in text classification can be applied to images to achieve better classification performance than Bag of Visual Words methods as it results in more complete image representation. However, this adds to the dimensionality and computational cost. State of the art Deep learning models have been successful for image classification. However, huge training data required for these models is a big challenge. This book chapter reviews the literature on Bag of Visual Words and N-gram models for image classification and retrieval. It also discusses few cases where the N-gram models have outperformed or given comparable performance to the state of the art Deep Learning Models. The literature demonstrates that N-grams is a powerful and promising descriptor for image representation and is useful for various classification and retrieval applications.

5.1 INTRODUCTION

Techniques for the automated classification of images rely heavily on approaches that transform the image's digital encoding into features derived for the.[76] Low level features such as color, texture, and shape were proposed for this in the early 1990s.

Color features such as histograms were common for image classification Texture is nothing but information about arrangement of intensities. Various Texture features have been used for classification of images. These are first order statistical features which are not able to provide much information about spatial correspondence.[73] Second order statistical features such as based on co-occurrence matrix based features are found to be powerful in distinguishing among various texture images.[1] Another texture feature used is Local Binary Pattern (LBP), which is sensitive to noise in uniform regions.[78] Spectral methods such as Gabor Filters,[30] Fourier Transforms,[60] which converts the image into signal by sampling were also popular for image classification tasks. Shape features such as Zernike Moments[61] were also been used for classification tasks. However, they are computationally expensive.

However, image representation using low level features suffered from the semantic gap problem. The semantic gap[12] is the gap between the high level concepts (for example, "Find pictures of Sunset") expected in a user's query and the information modeled by low level features. Moreover, it is difficult for a user to search for images using criteria such as color, texture, and shape.[25] Most importantly, low level features are the global image features and represent the image as a whole, but do not give much information about local pattern variations.

The idea of capturing local pattern variations in an image gave rise to the use of Bag-of-Visual-Words models (BoVW) for image representation.[74] BoVW model was inspired by Bag-of-Words (BoW) model in the text retrieval domain which has been proven to be efficient and is now widely deployed.[87] Text documents mainly contain meaningful words and so can be represented by a feature vector of counts of various words appearing in the document. A BoVW approach was first applied to video retrieval by Sivic and Zisserman.[74] In this approach, an image is described by a number of occurrences of different visual words. Visual words are local image patterns, which can describe relevant semantic information about an image. This model soon became popular for image retrieval and classification applications due to its accuracy.[11,32,59,66,68,82]

However, there are fundamental differences between text and images. First, text words are discrete tokens whereas local image descriptors are not. This necessitates techniques to generate a visual vocabulary by clustering the local feature descriptors. Vector quantization is a common technique for this but, in contrast to the text BoW, the feature vector generated is typically highly dimensional and the generation process is computationally complex.[81] Second, text is unidirectional whereas images can be read in several different directions.

Although, the BoVW model has proven to be much better than models using low level features such as color and texture,[80] it has major drawbacks. The BoVW model does not consider spatial relationships among visual words. Another BoVW drawback involves the high computational cost to generate vocabularies from low level features.[79] Further, the vocabulary construction process often results in noisy words that diminish classification.[79]

The visual N-grams model was first proposed for images. In order to take spatial relations between visual words into account.[62] There are two types of N-grams formulations in the text retrieval context. Word N-grams are formed by sequences of N consecutive words in a document; whereas, character N-grams are formed by sequence of N consecutive characters. Examples of word 2-gram are "image processing," "artificial intelligence," "medical systems," etc. In contrast, examples of character N-grams are the 3-grams in the phrase "his pool" "his, is_,s_p, _po, poo,ool" and the 4-grams "his_,is_p, s_po, _poo, pool." The N-gram model had proven to be more accurate than other models in text context.[54] Therefore, its application for image classification by Pedrosa and Traina,[62] promised semantically meaningful image representation. Since then, visual N-grams for images have not been widely researched despite favorable early results. In addition, pixel N-grams inspired from the character N-grams have also only recently been advanced.

This chapter provides a detailed information of Visual N-grams in relation to BoVW models for image classification and retrieval applications so that the differences between these two approaches can be clearly described. It also discusses the work using BoVW and N-gram approaches which have outperformed the state-of-the art deep learning approaches. This book chapter is organized as follows: Section 5.2 describes local features used for constructing image models for BoVW and N-grams. Section 5.3 describes the vocabulary/dictionary creation to highlight the source of the computational complexity. Section 5.4 outlines various approaches for visual N-gram generation to analyze claims of computational efficiencies. Some of the major challenges of BoVW and N-gram approaches are discussed in Section 5.5. Section 5.6 mentions various deep learning approaches for image classification. Section 5.7 concludes the paper.

5.2 LOCAL FEATURE EXTRACTION FOR BOVW AND N-GRAMS

The BoW model was first introduced in the text retrieval and categorization domain where a document is described by a set of keywords and their frequency of occurrence in the document. The same idea was applied to the image domain and has been quite successful.[74] Here, the idea is to represent an image using a dictionary of different visual words. Images are quite different from text documents in the sense that there is no natural concept of a word in case of images.[4] Thus, there is a need to break down the image into a list of visual elements. Moreover, as the number of possible visual elements in an image could be enormous, these elements should be discretized to form a visual word dictionary known as a codebook.

Vocabulary construction has been achieved mainly using two approaches: local, patch-based approach or dense sampling[4,48] and key point-based approach or sparse sampling.[16,62,65] In the patch-based approach, the image is divided into a number of equal sized patches by using a grid. Local features are then computed for each patch separately. Keypoints are the centers of salient patches generally located around the corners and edges. Keypoints are also known as interest points and can be detected using various region detectors such as the Harris–Laplace detector (corner-like structures), Hessian-affine detector,[79] Maximally stable extremal regions or the Salient regions detector.[55] Local features are then computed for each interest point.

Some of the state-of-art local feature descriptors used for modeling texture information include Scale Invariant Feature Transform (SIFT),[53] Speeded Up Robust Features (SURF),[5] Histogram of Oriented Edges (HOG),[18] Local Ternary Pattern (LTP),[78] and Discrete Cosine Transform (DCT).[15] Color hues and shape features have also been used as local feature descriptors by some of the researchers. These local feature descriptors are briefly described below.

SIFT descriptors[53] are invariant to image translation, illumination, noise, scaling, rotation, and partially invariant to illumination changes. These features are robust to local geometric distortion[55] and are the most commonly used local feature descriptors for the BoVW model. However, the limitations include high computational cost and the huge feature vector dimension (128 dimensions for each keypoint).

SURF features[5] are modified SIFT features. SURF features are high-performance, scale and rotation-invariant and they outperform SIFT features with respect to repeatability, distinctiveness, and robustness.[34] Computation time for calculating SURF features is reduced with the use of a fast Hessian matrix-based detector and a distribution-based descriptor. SURF descriptors have been successfully applied for diabetic retinopathy lesion detection,[33] video stabilization,[67] video copy detection,[89] recognition of museum objects,[6] and multi-person tracker.[20]

HOG[18] is also a simplified form of SIFT. It differs from SIFT in which it is computed on a dense grid of uniformly spaced cells and uses overlapping local contrast normalization. It calculates intensity gradients from pixel to pixel and selects a corresponding histogram bin based on gradient direction. The key advantage of HOG is that it is invariant to geometric and photometric transformations and is more accurate as compared to wavelets as well as SIFT[85] for human detection and scene categorization. However, HOG descriptors are dependent on the angle of the acquisition camera.[14]

Another popular feature used for capturing texture information are LTP.[78] These features are calculated using the binary difference between Gray value of a pixel and Gray values of P neighboring pixels on a circle of radius around it.[78] They have been used for different applications such as texture classification, face recognition, and background subtraction in complex scenes.[50] Advantages of LTP include rotation invariance and less sensitivity to noise as the small pixel difference is encoded into a separate state. To reduce the dimensionality, the ternary code is split into two binary codes: a positive LBP and a negative LBP. However, this splitting may cause significant information loss.

Features using a spectral approach (frequency domain) such as Discrete Cosine Transform (DCT) are also used for image classification applications. Some of these applications include histology image classification,[15] detecting pornographic video content[31]; object classification[19]; HE-p2 image classification[72]; histopathology image classification.[52] These features outperformed SIFT descriptors.[15] An advantage of DCT is the capacity to pack the energy of spatial sequences into as few frequency coefficients as possible known as energy compaction thus reducing the feature vector dimensionality. The main disadvantage of DCT is the blocking effect. When the image is reduced with higher compression ratios, the blocks become visible degrading the picture quality.

The features described above are texture features and are mainly based on intensity distribution of the pixels in an image. Apart from texture, color is also a powerful image descriptor. Use of Color Hues (Hue, Saturation and Intensity) along with the N-gram concept was first proposed by.[71] This technique preserves some of the spatial color correlates within an image to provide a more selective matching mechanism than global color histograms. Here, images are encoded with respect to a codebook of features which describes every possible combination of a fixed number of coarsely quantized color hues that might be encountered within local regions of an image. This enables images to be compared on the basis of their shared adjacent color artifacts or boundaries. This approach is analogous to a technique employed in text retrieval systems which use character substrings as the basis of the indexing and matching mechanism.[77]

Shape is the another important property for representation of an image. Various shape features such as Zernike moments, wavelet transforms, and Bsplines have been widely used for image representation. These features are based on mathematical formulations and have little to do with human visual perception. Perceptual features are higher level representations which try to capture richer semantic content and exploit human visual perception rules. The idea of perceptual shape features was used by Mukanova et al.,[56] for creating shape N-grams. These N-gram-based perceptual shape features can efficiently represent global shape information in an image and are seen to significantly increase the performance of the SIFT-based BoVW approach. The main drawbacks of the shape features are the computational cost and the requirement for segmentation.

Two main sampling strategies are employed in order to compute the above-mentioned local features. The SIFT, SURF, and Shape-based features are considered to be sparse descriptors or keypoint-based descriptors; whereas the HOG, LTP, DCT, Color Hues are considered dense descriptors or patch-based descriptors. It is clear that the dense sampling descriptors outperform the sparse descriptors as some of the information is lost in keypoint-based approach.[7,35]

5.3 VOCABULARY CONSTRUCTION

After calculation of local features, the next step in the BoVW or N-gram representation of an image is the vocabulary construction. Since an image does not contain discrete visual words, a challenging task is to discover meaningful visual words. This can be achieved by clustering local features so that cluster centroids can be treated as visual words. Various clustering algorithms such as Generalized Llyod Algorithm (GLA), Pairwise Nearest Neighbor Algorithm (PNNA) and K-means Algorithm have been widely used for this purpose.[92] However, GLA is computationally complex and cannot guarantee an optimal codebook generation.[92] On the other hand, PNNA is more efficient than GLA but slightly inferior to GLA in terms of optimality.[91,92] Further, the K-means algorithm performs better than the hierarchical algorithms in terms of accuracy and computation time. It differs from the GLA in that the input for k-means algorithm is the discrete set of points rather than continuous geometric region. This algorithm partitions N number of local features into K clusters in which each feature belongs to the cluster with the nearest mean. This is the most commonly used algorithm for visual codebook generation.[4,8,16,46,52,58,65,75,83,90]

The approaches for vocabulary construction can be mainly grouped under two main categories: global dictionary and sub-dictionary. If a single dictionary of visual words is created using all the images in the collection, it is called as global dictionary.[4,18,52,56,62,75,79,90,91] On the contrary, sub-dictionary approach considers subset of visual words that best represent a specific image class and is also known as region-specific visual words. For example, in diabetic retinopathy images, two sub-dictionaries related to lesion and no-lesion classes can be separately created.[33] Classification as well as retrieval performance can be improved over the global dictionary approach using the sub-dictionary approach.[29,64]

Creation of visual N-gram codebook can be more challenging than the BoVW codebook creation. This is due to the fact that as opposed to text, an image can be read in many different directions (horizontal, vertical, at an angle of Θ degrees). Further, visual N-grams that have the same order but different orientations may be related to the same pattern. One such approach of generating rotation invariant N-gram codebooks can be seen in the work of López-Monroy et al.[52] Moreover, as N increases the dictionary size is increased tremendously if we consider all possible combinations of visual words in all possible directions.

5.4 DIFFERENT APPROACHES FOR VISUAL N-GRAMS

Although, BoVW generates promising results in image retrieval and classification tasks; loss of spatial information and noisy words creation are two major drawbacks of this approach.[79] The limitation of spatial information loss could be overcome by using visual N-grams.[90] N-grams is a description obtained by grouping visual words where the arrangement between the visual words in an image is encoded. This is because the appearance of the visual words can change profoundly when they participate in relations. Further, the N-gram models for image features are simple and are able to scale up the content representation just by increasing N.[62]

By analogy to the text document (see Figure 5.1), there are mainly two approaches for visual N-grams image representation. Visual Word N-grams model the spatial relationship among the visual words. In contrast, Visual Character or Pixel N-grams, model the relationships among the pixels in various directions. These approaches are detailed below.

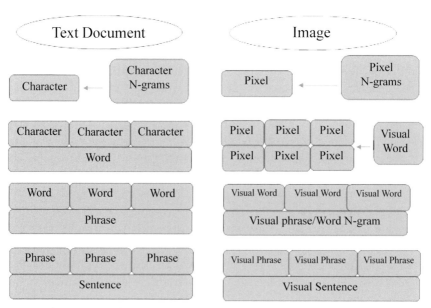

FIGURE 5.1 Text and image N-gram analogy.

5.4.1 KEYPOINT-BASED N-GRAMS

Keypoints/Interest points are the points of local maxima and minima of difference of Gaussian function.[53] These keypoints are described with the help of SIFT features and clustered for construction of visual vocabulary. The centroids of the clusters represent visual words. The N-gram dictionary is then created considering N neighboring visual words in all possible directions.[62] It is evident that as N increases, a more complete representation of an image is generated. Here, authors have used 1, 2, and 3 g to analyze retrieval precision as well as classification accuracy on various databases namely Corel 1000,[47] Lung database, Medical Image Exams database, Texture database. These experiments show that the visual word N-grams (bag-of-visual-phrases) approach improved retrieval precision up to 44% and classification accuracy up to 33%. However, the use of visual words to represent an image in this way may involve a loss of fidelity to visual content since two local features associated with the same visual word are used in the same way to construct the image signature, whether they are identical or noticeably different. An approach for generating a more realistic image signature considering the differences between textual words and visual words can be seen in the work of some researchers.[8,9] Some more examples of the use of keypoint-based N-grams are large scale image retrieval,[17] automatic learning of visual phrases,[83] classification of images in Caltech dataset[56] and biomedical image classification.[52,63,64]

For visual characterization, the frequency of occurrence of visual words as well as the spatial information between the visual words is equally important. A major challenge in using word N-grams is the dimensionality and hence the computational cost. It is clear that the number of all possible combinations of N-grams increases exponentially with N. That is, given a dictionary with m words, the number of all possible N-grams is m^N.

A novel effective and efficient technique to extract the frequency and appearance of visual words has been proposed in Pedrosa et al.[63] In this approach, 2-grams are generated by placing a circular region over each keypoint. All pairs of words in this region formed with the centre point are 2-grams. Two bags of 2-grams are then generated. One bag for 2-grams with angle within [−135, 135] and [−45, 45] and another bag for 2-grams with angle within the interval [135, 45] and [−135, −45]. Then the frequency of 2-grams for each bag according to dictionary of 2-grams is noted. This is called as bag-of-2-grams approach. The results demonstrate that the classification accuracy is improved by 6.03% as compared to the BoVW approach. Further, this approach computes the Shannon entropy over a random "bunch" of 2-grams and demonstrates that the dimensionality can be significantly reduced.

Keypoint-based approaches identify points throughout the image that are used as reference points from which N-grams are generated in one way or another. Local patch-based N-grams, discussed next replace the keypoints, with regions called keyblocks.

5.4.2 LOCAL PATCH-BASED N-GRAMS

In this approach, an image is divided into small local patches using a grid. Local features are computed for each patch separately. A codebook or dictionary of visual words is then created by clustering all the patch descriptors. N-gram codebook is then developed by considering the N-consecutive visual words present in an image.

The idea of N-grams using local patches was first proposed by Zhu et al.,[91] and was called the keyblock approach. Keyblocks are similar to key words in text and using these keyblocks, images can be represented as a code matrix in which the elements are indices of keyblocks in the codebook. Uni-block, Bi-Block (horizontal, vertical, diagonal), and Tri-Block (horizontal, vertical, diagonal, triangular) configurations were used. The disadvantage of Bi- and Tri-Block models is increased dimension of feature vector requiring large storage and, therefore, less efficiency and retrieval performance because of highly sparse nature. However, the dimensionality can be reduced by selecting only useful Bi- and Tri-Blocks. It is reported that combination of Uni-, Bi-, and Tri-blocks result in improvement in retrieval performance. Experiments were conducted on Brodatz texture database (TDB),[10] CDB (snapshot of images on web). Keyblock approach is compared with traditional color histogram and color coherent vector techniques using CDB and compared against Haar and Daubechies wavelet texture techniques using TDB. Using the keyblock approach, 12% of all relevant images are among top 100 retrieved images as compared to 9% of color histogram and 6.5% returned by Color Coherent Vector. Also, at each recall level keyblock approach achieved higher precision. In this study, it has also been observed that the keyblock approach outperforms the Haar and Daubechies wavelet texture approaches.

Recently, local patch-based N-grams were used for histopathological image classification.[52] The local patches were represented using DCT features. Here, the main idea was to produce N-grams ignoring the orientation in which they appear. Visual N-grams that have the same order but different orientation (e.g., if an image is rotated), like 12-65-654 and 654-65-12 are considered same, thus making the N-gram features rotation invariant. Another main idea in this study was to combine the N-gram features such as 1 + 2 gram, 1 + 2 + 3 gram and 1 + 2 + 3 + 4 gram. The 1 + 2 gram produced the highest classification accuracy of 64.31%. The reason is because longer sequences produce large vocabulary resulting in sparse feature vector. Results re-enforce the fact that the use of N-grams outperform the BoVW technique. Composing simple image descriptions using the patch-based N-grams can be seen in Li et al.[49] It is observed that keypoint-based samplers such as Harris–Laplace work well for small numbers of sampled patches; however, they cannot compete with uniform random patch-based sampling using larger numbers of patches for best classification results.[35]

5.4.3 COLOR N-GRAMS

Color features have been used for CBIR because they can be easily extracted and are powerful descriptors for images. Color histograms representing relative frequency of color pixels across the image are common for CBIR. However, they only convey global image properties and do not represent local color information. In the Color N-grams approach, an image has been represented with respect to a codebook, which describes every possible combination of a fixed number of coarsely quantized color hues.[71] This allows comparison of images based on shared adjacent color objects or boundaries. N-gram samples were taken to be 25% of the total number of pixels in an image. The dataset included 100 general color images of faces, flowers, animals, cars, and aeroplanes. The results were compared with the approach adopted by Faloutsos et al.[21] The average rank of all relevant images was reported to be 2.4 as compared to the 2.5 of the baseline. Also the number of relevant images missed was 1.9 as compared to 2.1 of the baseline. The limitation of this study is that the quantization of the hues does not match the sensitivity of the human color perception model. Another limitation was the very small database used. However, further work has demonstrated that this approach could also be used for very large databases.[70] Moreover, this approach is less sensitive to small spectral differences and is not prone to color constancy problems.

5.4.4 SHAPE N-GRAMS

The concept of N-gram has been used to group perceptual shape features to discover higher level semantic representation of an image.[56] Here, low-level shape features are extracted and perceptually grouped using the Order Preserving Arctangent Bin (OPABS) algorithm advanced by Hu and Gao. This is based on perceptual curve partitioning and grouping PCPG model.[23] In this PCPG model, each curve is made up of Generic Edge Tokens (GET) connected at Curve Partitioning Points (CPP). Each GET is characterized by monotonic characteristics of its Tangent Function (TF) set. The extracted perceptual shape descriptors are categorized as one of eight generic edge segments.

Gao and Wang's model is based on Gestalt's theory of perceptual organization which states that humans perceive the objects as a whole. The authors define shape N-gram as continuous subsequence of GETs connected at CPP points. There are three main cases of how the GETs are connected at CPP. The first references a curve segments connected to another curve segment (CS–CS); the second is a line segment connected to line segment (LS–LS), and the third is curve segment connected to line segment (CS–LS). Here, four N-gram based perceptual feature vector are proposed, which encode local and global shape information in an image. The Caltech256 dataset was used for classification experiments.[27] Results show that the combination of shape N-grams with conventional SIFT vocabulary achieve around 8% higher classification accuracy as compared to SIFT-based vocabulary alone.

Further, the development of CANDID (Comparison Algorithm for Navigating Digital Image Database)[37] was inspired by the N-gram approach to document fingerprinting. Here, a global signature is derived from various image features such as localized texture, shape, or color information. A distance between probability density functions of feature vectors is used to compare the image signatures. Global feature vectors represent single measurement over the entire image (e.g., dominant color, texture). Whereas, the N-gram approach allows for the retention of information about the relative occurrences of local features such as color, gray scale intensity or shape. Use of probability density functions can reduce the problem of high dimensions; however, they are computationally more expensive than histogram-based features.[38] It is observed that subtracting a dominant background from every signature prior to comparison does not have any effect while using true distance function; whereas, considering a similarity measure such as nSim(I1,I2), dominant background subtraction has a dramatic effect. The experiments were conducted on satellite data (LandSet TM 100 images) and Pulmonary CT imagery (220 lung images from 34 patients). Experimental results show good retrieval precision.

The word N-gram approaches are divided into keypoint based and local patch based according to the sampling strategies used; whereas, based on local features used these approaches are divided into color N-grams and shape N-grams. Another concept called character N-grams in the text retrieval domain has also been applied recently for image representation. This is described below.

5.4.5 VISUAL CHARACTER/PIXEL N-GRAMS

In the text, retrieval context Character N-grams are phrases formed by N consecutive characters. For languages such as Chinese, where there are no specific word boundaries, the character N-grams have resulted in higher retrieval accuracies and are found more efficient than the word N-gram model in several cases.[36] If we consider every pixel in an image as a character, the character N-gram concept from text retrieval can be easily applied to the image representation. A first attempt to apply the character N-gram concept for mammographic image classification show promising results.[41,42] It has been observed from the further experiments that the visual character N-grams (Pixel N-grams) outperform the traditional co-occurrence matrix-based features for classification of mammograms. Moreover, the character N-gram features are found to be seven times faster than the co-occurrence matrix feature computation.[43] The Pixel N-grams also show improved classification performance compared with the BoVW for texture classification experiment with an added advantage of simplicity and less computational cost.[44] Thus the Pixel N-grams try to overcome the two drawbacks of the visual word N-grams; namely computational complexity and feature vector dimensionality.

5.4.6 VISUAL SENTENCE APPROACH

A new representation of images that goes further in the analogy with textual data, called visual sentences, has been proposed by Tirilly et al.[79] A visual sentence that allows visual words to be read visual words in a certain order. An axis is chosen for representing an image as a visual sentence, so that (a) it is at an orientation fitting the orientation of the object in the image, (b) it is at a direction fitting the direction of the object. The keypoints are then projected onto this axis using orthogonal projection. In this work, SIFT descriptors are used and keypoints detection is achieved using Hessian-affine detector. The main problem is to decide the best axis for projection. Experiments include five different axis configurations: 1 PCA axis, 2 orthogonal PCA axis, 10 axis obtained by successive rotation of 10 degrees of main PCA axis, X-axis and finally one random axis. Results show that the approach with X-axis outperforms those with the PCA axis on classification tasks.[13,40,69] This is because the PCA axis is biased by background clutter. However, PCA axis takes spatial relations into account and outperforms the random axis or the multiple axis configurations.[70,86]

5.4.7 CONTEXTUAL BAG-OF-WORDS

Two relations between local patches in images or video keyframes can be important for categorization. First, there is the semantic conceptual relation between patches. That is the relation of appearing on the "same part," "same object," or "same category." For example, "wheel of a motorbike," "window of a house," "eye of human." Further, semantic relations can be interpreted in multiple levels, for example, patches of same scene, object, object parts, and so on. Second is the spatial neighborhood relation. Patches when combined together to form a meaningful object or object part are considered as having spatial neighborhood relation. These two types of relations are called as contextual relations. Traditional BoW model neglect the contextual relations between local patches. Nevertheless, it is well known that the contextual relations play an important role in recognizing visual categories from their local appearance. A contextual-bag-of-words (CBOW) considers two types of relationships between local patches. On the 15 scene database, the classification accuracy using CBOW is found to be significantly better than the traditional BoVW model.[49]

The major problem of BoVW or N-grams approach is the feature vector dimension and, hence, computational cost. Various ways to achieve dimension reduction are discussed in the next section.

5.5 CHALLENGES OF VISUAL N-GRAM APPROACHES

Table 5.1 displays a summary of various approaches of N-grams for image classification and retrieval applications. Despite the early success of BoVW and N-grams model with regards to classification and retrieval performance, the use of these models face few critical challenges.

One of the major challenges in representation of images using BoVW and visual N-gram model is the construction of the codebook or visual words dictionary. Mainly two types of codebook generation can be observed. The first is a global dictionary where the patches or keypoints in the entire image are clustered for creating a single dictionary (Avni et al., 2010). The second approach uses sub-dictionaries or region specific BoW (Jelinek et al., 2013; Pedrosa et al., 2014; Wei Yang et al., 2012). Using sub-dictionaries has shown to boost the mAP by 2% and classification accuracy by 4.25% as compared to global dictionary representation (Pedrosa et al., 2014). Furthermore, various clustering algorithms play an important role in the visual codebook creation; for example, GLA, PNNA, and k-means algorithm out of which k-means is the most commonly used clustering algorithm due to its efficiency and optimality.

Another challenge is to reduce the dimensionality of the feature vector resulting in the reduction of computation cost. This can either be achieved by reducing the dictionary size or reducing the size of feature vector representing the local patch or keypoint. One way to reduce the dictionary size is by eliminating the visual words common to all the categories of images as they add very little discriminating power to the feature vector. Another way to reduce the dictionary size is the elimination of noisy words created due to the coarseness of the vocabulary construction process. These noisy words could be eliminated with the help of probabilistic latent semantic analysis. The dictionary size could also be reduced by ignoring the N-grams with repeated visual words and considering inverted N-gram as same visual phrase (Pedrosa and Traina, 2013). For reducing the dimensionality, an approach called "bunch of 2-grams" has been proposed by (Pedrosa et al., 2014). Here, the feature vectors are grouped in bunches

TABLE 5.1 Summary of Various N-gram Approaches.

Author	Year	Model	Local features	Dataset	Advantages	Application
Kelly et al.	1994	CANDID	5 one dimensional kernels	Pulmonary CT scans	Features invariant to rotation	Retrieval pulmonary diseased cases
Rickman and Rosin	1996	Color N-grams	Color hues (Hue saturation and intensity)	100 color images	Robust to noise Rapid fuzzy matching of color images	Retrieval of color images
Soffer	1997	N×M grams	N × M grams absolute count and frequency	Fingerprint, floorplans, comics, animals etc.	Works well on simple images such as floorplan, music notes, comincs	Image categorization
Zhu et al.	2000	Key-block		CDB-500 web color images divided into 41 groups TDB-2240 Gray scale Brodatz texture images divided into 112 categories	Superior to color histogram, color coherent vector, Haar and Daubechies wavelet texture approach	Image retrieval
Zhu et al.	2002	n-block Bi-block and Tri-block		Corel :31646 images CDB: web color images Brodatz texture database	Superior to color histogram, color coherent vector, Haar and Daubechies wavelet texture features	Image Retrieval
Lazebnik et al.	2006	Pyramid matching	SIFT	Caltech-101 Graz	Improved performance than the orderless representation	Recognising natural scene categories
Zhang et al.	2006	Bag of Visual Phrases	SIFT (Salient local patch)	Caltech-101: 8707 images of 101 classes	BoVP approach is 20% more effective than BoVW	Retrieve images containing desired objects

TABLE 5.1 (Continued)

Author	Year	Model	Local features	Dataset	Advantages	Application
Wu et al.	2007	Extended document model	Raw pixel value SIFT Texture histogram	Caltech Corel	Robust to translation, illumination variance, view point change and complex backgrounds	Image Classification
Tirilly et al.	2008	Visual sentence	SIFT	Caltech-101	Visual sentences are independent of rotation and scaling	Image classification
Li et al.	2011	N-grams	objects, attributes spatial relationships	PASCAL2010	Web scale N-grams can be used to create sentences to annotate an image	Image annotation sentence
Li, Mei, Kwen, Hua	2011	Conceptual bag-of-words	SIFT (Dense Local patches)	TRECVID2005	Superior performance than convenient BoVW	Video event and scene categorization
Dai et al.[17]	2013	Visual groups	Master feature and member features	Oxford Buildings 5k (5062 images) Flickr 1M : Images of famous landmarks	Outperforms BoVW Model. Inclusion relationship is invariant to image transformations	Large scale image retrieval
Pedrosa et al.	2013	BoVP	SIFT	Corel 1000: 1000 images Lung CT: 234 images Medical exams database: 2200 x-ray and MRI	BoVP improves upto 44% of retrieval precision and 33% classification rate compared to BoVW	Image Classification and retrieval
Wang et al.	2013	BoVP	HOG, Shape context, SIFT	CTC data from 20 patients for teniae detection CTC data from 50 patients for polyp classification	Automatic way to learn visual phrases	Computer aided teniae detection, classification of colorectal polyp

TABLE 5.1 *(Continued)*

Author	Year	Model	Local features	Dataset	Advantages	Application
Battiato et al.	2013	N-grams	SIFT	Flickr: 3300 images UKBench: 10200 images	Exploit coherence between feature space not only in image representation step but also during codebook creation. Outperforms BoVP	Near duplicate image detection
Monroy et al.	2013	N-gram combination	Discrete cosine transform	Histopathological dataset: 1417 images of 7 categories	1 + 2 grams improves accuracy by 6% than BoVW	Histopathological classification for basal cell carcinoma
Mukanova et al.	2014	Shape N-grams	SIFT, perceptual shape features	Wang: 100 images of 10 categories Caltech 256: 10 classes each with 80 images	Improves accuracy by 8% as compared to traditional BoVW	Classification of images
Pedrosa et al.	2014	Bag-of-2-grams Bunch of 2-grams	SIFT	ImageCLEFmed 2007: 5042 biomedical images of 32 categories	Classification accuracy is improved by 6.03% as compared to traditional BoVW	Biomedical image classification
Pedrosa, Triana	2014	Sub-dictionaries	SIFT	ImageCLEF 2007: 5042 images of 32 categories	Boosted mAP by 2% and classification accuracy by 4.25% as compared to BoVW.	Image retrieval and classification
Ruber Hernández-García et al.	2018	N-gram + Graph	SIFT	KTH Weizmann UCF Sports UCF Youtube Hollywood2	Improvement in mean average precision is noticed as compared to BoVW	Action recognition from videos

and each bunch is represented using its Shannon entropy. A dimensionality reduction up to 99% can be achieved using this approach.

A problem with the word N-gram approach is that the number of phrases can exponentially grow with respect to number of words in a phrase. A subset of these phrases may be selected by using sophisticated mining algorithms, but it is still risky to discard a large number of phrases as some of which may be the representative ones for an image. Moreover, with increase in N, this model produces very specific features which make it difficult for classifiers to generalize well. For this model, the computational cost can be further reduced by gray scale reduction, but choosing a gray scale reduction to preserve image details and increase noise robustness while reducing the computational cost is a challenging task. Also, the choice of N for best performance varies according to the dataset used.

5.6 DEEP LEARNING MODELS FOR IMAGE CLASSIFICATION

Deep learning neural networks are part of machine learning algorithms based on artificial neural networks. These networks basically do not require handcrafted features. Various architectures of deep learning techniques exist, such as convolution neural networks, deep neural networks, recurrent neural networks, and deep belief networks.

Breast cancer detection using deep learning framework was quite successful for cytology images.[39] Combination of Convolution Neural Network (CNN) and extreme machine learning achieved 99.5% accuracy for classifying cervical cancer detection.[24] CNNs have also been shown to achieve high accuracy for brain tumor classification by Amin et al..[2] The editorial discusses how the deep learning methods are used for medical image segmentation, computer aided detection, classification tasks.[26]

Bag of Visual Words and N-grams are still a good option in case of less training data. A study by Kumar[45] shows that the BoVW (96.5% accuracy) has worked better than the deep learning models (94.76% accuracy) for histopathological image analysis. Another work by Huang demonstrates that N-gram applied to focal liver lesions classification using CT images have provided 83% accuracy and also high training speed.[28,29] Lopez-Monroy used the Distributional Term Representations (DTR) for various image datasets and has demonstrated that this technique works better than the deep learning neural networks.[51] The DTR technique is modified form of visual N-grams and considers statistics of visual word occurrences and co-occurrences.

Although, deep learning has several advantages over traditional methods it becomes challenging in the field of medical image classification because of the significant intra-class variation and inter-class similarity caused by the diversity of clinical pathologies and imaging modalities. A synergic deep learning approach has been proposed by Zhang et al.,[88] to overcome this limitation. Another disadvantage using deep learning networks is that the network requires a lot of training data and getting annotated medical images for training is a big challenge.[57,84,86] Deep learning is computationally very expensive and requires high power computational resources such as GPU environment and lot of training time.[3] It is very difficult to comprehend what is learnt by the deep neural network.[36,42]

BoVW model performance is found less satisfactory for geographical imagery because of the complexity and diversity of landscape. In this case, an experiment combining the CNN-based spatial features and the BoVW-based image interpretation was very much successful for geographical image classification task.[22]

5.7 CONCLUSION AND FUTURE DIRECTIONS

In this chapter, we discussed the literature on BoVW and N-gram models for image classification and retrieval applications with respect to local features used, vocabulary construction process and various N-gram approaches. It is evident that the BoVW model gives better classification and retrieval performance as compared to global image features such as texture, color, and shape. BoVW model, however, does not incorporate spatial relationships. N-gram model try to incorporate the spatial relationship and increase performance but certainly add to the computational complexity. N-grams can be classified based on sampling strategies (sparse and dense). Dense sampling approaches work better than the sparse sampling. We looked at various N-gram approaches namely keypoint based, local patch based, color based, shape based, Pixel based, visual sentence approach, and CBOW. The literature demonstrates that N-grams is a powerful and promising descriptor for image representation and is useful for various applications such as content-based image retrieval, classification, annotation, action recognition for various types of images (natural scenes, biomedical images, texture images, fingerprints) and videos.[13,17,40,69]

However, some of the challenges of N-gram models is to reduce the vocabulary size, computational cost, choice of sampling strategy, choice of local features, choice of clustering algorithm, dimension reduction and reducing noisy word creation during vocabulary construction process.

Recent trend for image classification/retrieval is to use deep learning neural networks. In deep learning approach, the work of generating and optimizing image features is automatically done by the various layers of deep neural networks. We have discussed some literature where the visual N-grams model has outperformed deep learning models with an added advantage of less training time and less amount of training data. Finally, we have also shown the experiments where the features of BoVW and features from deep learning can be combined for better accuracy of large dataset. Therefore, we conclude that the visual N-grams are still a good choice for many image classification and retrieval applications where the datasets are small.

KEYWORDS

- **content based image retrieval (CBIR)**
- **N-grams**
- **image classification**
- **Local Binary Pattern (LBP)**
- **Pairwise Nearest Neighbour Algorithm (PNNA)**

REFERENCES

1. Aggarwal, N.; Agrawal, R. First and Second Order Statistics Features for Classification of Magnetic Resonance Brain Images. *J. Sign. Info. Process.* **2012,** *3* (2).
2. Amin, J.; Sharif, M.; Gul, N.; Yasmin, M.; Shad, S. A. Brain Tumor Classification Based on DWT Fusion of MRI Sequences Using Convolutional Neural Network. *Pattern Recogn. Lett.* **2020,** *129*, 115–122.
3. Angelov, P.; Sperduti, A. *Challenges in Deep Learning.* Paper presented at the ESANN, 2016.

4. Avni, U.; Goldberger, J.; Sharon, M.; Konen, E.; Greenspan, H. *Chest X-ray Characterization: From Organ Identification to Pathology Categorization.* Paper presented at the Proceedings of the international conference on Multimedia information retrieval, 2010.
5. Bay, H.; Ess, A.; Tuytelaars, T.; Van Gool, L. Speeded-up Robust Features (SURF). *Comput. Vision Image Understand.* **2008,** *110* (3), 346–359.
6. Bay, H.; Fasel, B.; Gool, L. V. *Interactive Museum Guide: Fast and Robust Recognition of Museum Objects.* Paper presented at the Proceedings of the first international workshop on mobile vision, 2006.
7. Bosch, A.; Zisserman, A.; Muñoz, X. Scene Classification via pLSA. *Comput. Vision–ECCV 2006* **2006,** 517–530.
8. Samantaray, S.; Deotale, R.; Chowdhary, C. L.. Lane Detection Using Sliding Window for Intelligent Ground Vehicle Challenge. In *Innovative Data Communication Technologies and Application,* Springer: Singapore, 2021; pp. 871-881.
9. Bouachir, W.; Kardouchi, M.; Belacel, N. *Improving Bag of Visual Words Image Retrieval: A Fuzzy Weighting Scheme for Efficient Indexation.* Paper presented at the Signal-Image Technology & Internet-Based Systems (SITIS), 2009 Fifth International Conference on, 2009b.
10. Brodatz, P.; Textures, A. A Photographic Album for Artists and Designers. 1966. *Images downloaded in July,* 2009.
11. Caicedo, J. C.; Cruz, A.; Gonzalez, F. A. Histopathology Image Classification Using Bag of Features and Kernel Functions. *Artif. Intell. Med.* **2009,** 126–135.
12. Chen, Y.; Wang, J. Z.; Krovetz, R. *An Unsupervised Learning Approach to Content-based Image Retrieval.* Paper presented at the Signal Processing and Its Applications, 2003. Proceedings. Seventh International Symposium on, 2003.
13. Chowdhary, C. L. 3D Object Recognition System Based on Local Shape Descriptors and Depth Data Analysis. *Rec. Patents Comput. Sci.* **2019,** *12* (1), 18–24.
14. Climer, J. *Overcoming Pose Limitations of a Skin-Cued Histograms of Oriented Gradients Dismount Detector Through Contextual Use of Skin Islands and Multiple Support Vector Machines,* 2011.
15. Cruz-Roa, A.; Díaz, G.; Romero, E.; González, F. A. Automatic Annotation of Histopathological Images Using a Latent Topic Model Based on Non-negative Matrix Factorization. *J. Pathol. Info.* **2011,** *2*.
16. Csurka, G.; Dance, C.; Fan, L.; Willamowski, J.; Bray, C. *Visual Categorization with Bags of Keypoints.* Paper presented at the Workshop on statistical learning in computer vision, ECCV, 2004.
17. Dai, L.; Sun, X.; Wu, F.; Yu, N. *Large Scale Image Retrieval with Visual Groups.* Paper presented at the Image Processing (ICIP), 2013 20th IEEE International Conference on, 2013.
18. Dalal, N.; Triggs, B. *Histograms of Oriented Gradients for Human Detection.* Paper presented at the Computer Vision and Pattern Recognition, 2005. CVPR 2005. IEEE Computer Society Conference on, 2005.
19. Deselaers, T.; Ferrari, V. *Global and Efficient Self-similarity for Object Classification and Detection.* Paper presented at the Computer Vision and Pattern Recognition (CVPR), 2010 IEEE Conference on, 2010.

20. Ess, A.; Leibe, B.; Schindler, K.; Gool, L. V. *A Mobile Vision System for Robust Multi-person Tracking.* Paper presented at the Computer Vision and Pattern Recognition, 2008. CVPR 2008. IEEE Conference on, 2008.
21. Faloutsos, C.; Barber, R.; Flickner, M.; Hafner, J.; Niblack, W.; Petkovic, D.; Equitz, W. Efficient and Effective Querying by Image Content. *J. Intell. Info. Syst.* 1994, *3* (3–4), 231–262.
22. Feng, J.; Liu, Y.; Wu, L. Bag of Visual Words Model with Deep Spatial Features for Geographical Scene Classification. *Comput. Intell. Neurosci.* **2017,** *2017*.
23. Gao, Q.-G.; Wong, A. Curve Detection Based on Perceptual Organization. *Pattern Recogn.* **1993,** *26* (7), 1039–1046.
24. Ghoneim, A.; Muhammad, G.; Hossain, M. S. Cervical Cancer Classification Using Convolutional Neural Networks and Extreme Learning Machines. *Future Gen. Comput. Syst.* **2020,** *102*, 643–649.
25. Goodrum, A.; Spink, A. Image Searching on the Excite Web Search Engine. *Info. Process. Manage.* **2001,** *37* (2), 295–311.
26. Greenspan, H.; Van Ginneken, B.; Summers, R. M. Guest Editorial Deep Learning in Medical Imaging: Overview and Future Promise of an Exciting New Technique. *IEEE Trans. Med. Imag.* **2016,** *35* (5), 1153–1159.
27. Griffin, G.; Holub, A.; Perona, P. Caltech-256 Object Category Dataset, 2007.
28. Huang, H.; Ji, Z.; Lin, L.; Liao, Z.; Chen, Q.; Hu, H.; . . . Tong, R. Multiphase Focal Liver Lesions Classification with Combined N-gram and BoVW. In *Innovation in Medicine and Healthcare Systems, and Multimedia*; Springer, 2019; pp 81–91.
29. Huang, M.; Yang, W.; Yu, M.; Lu, Z.; Feng, Q.; Chen, W. Retrieval of Brain Tumors with Region-specific Bag-of-visual-words Representations in Contrast-enhanced MRI Images. *Comput. Math. Methods Med.* **2012,** *2012*, 280538. doi:10.1155/2012/280538
30. Hussain, M.; Khan, S.; Muhammad, G.; Berbar, M.; Bebis, G. *Mass Detection in Digital Mammograms Using Gabor Filter Bank.* Paper presented at the Image Processing (IPR 2012), IET Conference on, 2012.
31. Jansohn, C.; Ulges, A.; Breuel, T. M. *Detecting Pornographic Video Content by Combining Image Features with Motion Information.* Paper presented at the Proceedings of the 17th ACM international conference on Multimedia, 2009.
32. Jégou, H.; Douze, M.; Schmid, C. Improving Bag-of-features for Large Scale Image Search. *Int. J. Comput. Vision* **2010,** *87* (3), 316–336.
33. Jelinek, H. F.; Pires, R.; Padilha, R.; Goldenstein, S.; Wainer, J.; Rocha, A. *Quality Control and Multi-lesion Detection in Automated Retinopathy Classification Using a Visual Words Dictionary.* Paper presented at the Intl. Conference of the IEEE Engineering in Medicine and Biology Society, 2013.
34. Juan, L.; Gwun, O. A Comparison of Sift, PCA-sift and Surf. *Int. J. Image Process. (IJIP)* **2009,** *3* (4), 143–152.
35. Jurie, F.; Triggs, B. *Creating Efficient Codebooks for Visual Recognition.* Paper presented at the Computer Vision, 2005. ICCV 2005. Tenth IEEE International Conference on, 2005.
36. Kanaris, I.; Kanaris, K.; Houvardas, I.; Stamatatos, E. Words Versus Character N-grams for Anti-spam Filtering. *Int. J. Artif. Intell. Tools* **2007,** *16* (6), 1047–1067.
37. Kelly, P. M.; Cannon, T. M. *CANDID: Comparison Algorithm for Navigating Digital Image Databases.* Paper presented at the Scientific and Statistical Database

Management, 1994. Proceedings.; Seventh International Working Conference on, 1994.
38. Kelly, P. M.; Cannon, T. M.; Hush, D. R. *Query by Image Example: The Comparison Algorithm for Navigating Digital Image Databases (CANDID) Approach.* Paper presented at the IS&T/SPIE's Symposium on Electronic Imaging: Science & Technology, 1995.
39. Khan, S.; Islam, N.; Jan, Z.; Din, I. U.; Rodrigues, J. J. C. A Novel Deep Learning Based Framework for the Detection and Classification of Breast Cancer Using Transfer Learning. *Pattern Recogn. Lett.* **2019**, *125*, 1–6.
40. Khare, N.; Devan, P.; Chowdhary, C. L.; Bhattacharya, S.; Singh, G.; Singh, S.; Yoon, B. SMO-DNN: Spider Monkey Optimization and Deep Neural Network Hybrid Classifier Model for Intrusion Detection. *Electronics* **2020**, *9* (4), 692.
41. Kulkarni, P.; Stranieri, A.; Kulkarni, S.; Ugon, J.; Mittal, M. *Hybrid Technique Based on N-gram and Neural Networks for Classification of Mammographic Images.* Paper presented at the International Conference on Signal, Image Processing and Pattern Recognition, Sydney, Australia, 2014 Feb.
42. Kulkarni, P.; Stranieri, A.; Kulkarni, S.; Ugon, J.; Mittal, M. Visual Character N-grams for Classification and Retrieval of Radiological Images. *Int. J. Multimedia Its App.* **2014** Mar, *6* (2), 35.
43. Kulkarni, P.; Stranieri, A.; Kulkarni, S.; Ugon, J.; Mittal, M. *Analysis and Comparison of Co-occurrence Matrix and Pixel N-gram Features for Mammographic Images.* Paper presented at the International Conference on Communication and Computing, Banglore, India, 2015.
44. Kulkarni, P.; Stranieri, A.; Ugon, J. *Texture Image Classification Using Pixel N-grams.* Paper presented at the IEEE International Conference on Signal and Image Processing, Beijing, China, 2016 Aug.
45. Kumar, M. D.; Babaie, M.; Zhu, S.; Kalra, S.; Tizhoosh, H. R. *A Comparative Study of CNN, BOVW and LBP for Classification of Histopathological Images.* Paper presented at the 2017 IEEE Symposium Series on Computational Intelligence (SSCI), 2017.
46. Lazebnik, S.; Schmid, C.; Ponce, J. *Beyond Bags of Features: Spatial Pyramid Matching for Recognizing Natural Scene Categories.* Paper presented at the Computer Vision and Pattern Recognition, 2006 IEEE Computer Society Conference on, 2006.
47. Li, J.; Wang, J. Z. Automatic Linguistic Indexing of Pictures by a Statistical Modeling Approach. *IEEE Trans. Pattern Analysis Mach. Intell.* **2003**, *25* (9), 1075–1088.
48. Li, S.; Kulkarni, G.; Berg, T. L.; Berg, A. C.; Choi, Y. *Composing Simple Image Descriptions Using Web-scale n-grams.* Paper presented at the Proceedings of the Fifteenth Conference on Computational Natural Language Learning, 2011.
49. Li, T.; Mei, T.; Kweon, I.-S.; Hua, X.-S. Contextual Bag-of-words for Visual Categorization. *Circ. Syst. Video Technol., IEEE Trans.* **2011**, *21* (4), 381–392.
50. Liao, S.; Zhao, G.; Kellokumpu, V.; Pietikäinen, M.; Li, S. Z. *Modeling Pixel Process with Scale Invariant Local Patterns for Background Subtraction in Complex Scenes.* Paper presented at the Computer Vision and Pattern Recognition (CVPR), 2010 IEEE Conference on, 2010.
51. López-Monroy, A. P.; Montes-y-Gómez, M.; Escalante, H. J.; González, F. A. Novel Distributional Visual-Feature Representations for Image Classification. *Multimedia Tools App.* **2019**, *78* (9), 11313–11336.

52. López-Monroy, A. P.; Montes-y-Gómez, M.; Escalante, H. J.; Cruz-Roa, A.; González, F. A. *Bag-of-visual-ngrams for Histopathology Image Classification.* Paper presented at the IX International Seminar on Medical Information Processing and Analysis, 2013.
53. Lowe, D. G. Distinctive Image Features from Scale-invariant Keypoints. *Int. J. Comput. Vision* **2004,** *60* (2), 91–110.
54. Mcnamee, P.; Mayfield, J. Character n-gram Tokenization for European Language Text Retrieval. *Info. Retrieval* 2004, *7* (1–2), 73–97.
55. Mikolajczyk, K.; Tuytelaars, T.; Schmid, C.; Zisserman, A.; Matas, J.; Schaffalitzky, F.; . . . Van Gool, L. A Comparison of Affine Region Detectors. *Int. J. Comput. Vision* **2005,** *65* (1–2), 43–72.
56. Mukanova, A.; Hu, G.; Gao, Q. *N-Gram Based Image Representation and Classification Using Perceptual Shape Features.* Paper presented at the Computer and Robot Vision (CRV), 2014 Canadian Conference on, 2014.
57. Munappy, A.; Bosch, J.; Olsson, H. H.; Arpteg, A.; Brinne, B. *Data Management Challenges for Deep Learning.* Paper presented at the 2019 45th Euromicro Conference on Software Engineering and Advanced Applications (SEAA), 2019.
58. Nanni, L.; Lumini, A.; Brahnam, S. Ensemble of Different Local Descriptors, Codebook Generation Methods and Subwindow Configurations for Building a Reliable Computer Vision System. *J. King Saud Univ. -Sci.* **2014,** *26* (2), 89–100.
59. Nister, D.; Stewenius, H. *Scalable Recognition with a Vocabulary Tree.* Paper presented at the Computer vision and pattern recognition, 2006 IEEE computer society conference on, 2006.
60. Ojansivu, V.; Heikkilä, J. *Blur Insensitive Texture Classification Using Local Phase Quantization.* Paper presented at the International conference on image and signal processing, 2008.
61. Papakostas, G. A.; Boutalis, Y. S.; Karras, D. A.; Mertzios, B. G. A New Class of Zernike Moments for Computer Vision Applications. *Info. Sci.* **2007,** *177* (13), 2802–2819.
62. Pedrosa, G. V.; Traina, A. J. *From Bag-of-visual-words to Bag-of-visual-phrases Using n-Grams.* Paper presented at the Graphics, Patterns and Images (SIBGRAPI), 2013 26th SIBGRAPI-Conference on, 2013.
63. Pedrosa, G. V.; Rahman, M. M.; Antani, S. K.; Demner-Fushman, D.; Long, L. R.; Traina, A. J. *Integrating Visual Words as Bunch of N-grams for Effective Biomedical Image Classification.* Paper presented at the Applications of Computer Vision (WACV), 2014 IEEE Winter Conference on, 2014.
64. Pedrosa, G. V.; Traina, A. J.; Traina, C. *Using Sub-dictionaries for Image Representation Based on the Bag-of-visual-words Approach.* Paper presented at the Computer-Based Medical Systems (CBMS), 2014 IEEE 27th International Symposium on, 2014.
65. Pelka, O.; Friedrich, C. M. FHDO Biomedical Computer Science Group at Medical Classification Task Of Image CLEF 2015. *Working Notes of CLEF*, 2015.
66. Philbin, J.; Chum, O.; Isard, M.; Sivic, J.; Zisserman, A. *Object Retrieval with Large Vocabularies and Fast Spatial Matching.* Paper presented at the Computer Vision and Pattern Recognition, 2007. CVPR'07. IEEE Conference on, 2007.
67. Pinto, B.; Anurenjan, P. *Video Stabilization Using Speeded Up Robust Features.* Paper presented at the Communications and Signal Processing (ICCSP), 2011 International Conference on, 2011.

68. Rahman, M. M.; Antani, S. K.; Thoma, G. R. *Biomedical CBIR Using "bag of keypoints" in a Modified Inverted Index.* Paper presented at the Computer-Based Medical Systems (CBMS), 2011 24th International Symposium on, 2011.
69. Reddy, T.; RM, S. P.; Parimala, M.; Chowdhary, C. L.; Hakak, S.; Khan, W. Z. A Deep Neural Networks Based Model for Uninterrupted Marine Environment Monitoring. *Comput. Commun.*, 2020.
70. Rickman, R. M.; Stonham, T. J. *Content-based Image Retrieval Using Color Tuple Histograms.* Paper presented at the Electronic Imaging: Science & Technology, 1996.
71. Rickman, R.; Rosin, P. *Content-based Image Retrieval Using Colour N-grams.* Paper presented at the Intelligent Image Databases, IEE Colloquium on, 1996.
72. Shen, L.; Lin, J.; Wu, S.; Yu, S. HEp-2 Image Classification Using Intensity Order Pooling Based Features and Bag of Words. *Pattern Recogn.* **2014,** *47* (7), 2419–2427.
73. Sheshadri, H.; Kandaswamy, A. Experimental Investigation on Breast Tissue Classification Based on Statistical Feature Extraction of Mammograms. *Comput. Med. Imag. Graph.* **2007,** *31* (1), 46–48.
74. Sivic; Zisserman. Video Google: A Text Retrieval Approach to Object Matching in Videos; USA, 2003; pp 1470–1477.
75. Sivic, J.; Zisserman, A. *Video Google: A Text Retrieval Approach to Object Matching in Videos.* Paper presented at the Computer Vision, 2003. Proceedings. Ninth IEEE International Conference on, 2003.
76. Smeulders, A. W.; Worring, M.; Santini, S.; Gupta, A.; Jain, R. Content-based Image Retrieval at the End of the Early Years. *Pattern Analy. Mach. Intell., IEEE Trans.* **2000,** *22* (12), 1349–1380.
77. Suen, C. Y. N-gram Statistics for Natural Language Understanding and Text Processing. *Pattern Analy. Mach. Intell., IEEE Trans.* **1979,** *2,* 164–172.
78. Tan, X.; Triggs, B. Enhanced Local Texture Feature Sets for Face Recognition Under Difficult Lighting Conditions. *Image Process., IEEE Trans.* **2010,** *19* (6), 1635–1650.
79. Tirilly, P.; Claveau, V.; Gros, P. *Language Modeling for Bag-of-visual Words Image Categorization.* Paper presented at the Proceedings of the 2008 international conference on Content-based image and video retrieval, 2008.
80. Tsai, C.-F. Bag-of-Words Representation in Image Annotation: A Review. *ISRN Artif. Intell.* **2012,** *2012,* 1–19. doi:10.5402/2012/376804
81. van de Sande, K. E.; Gevers, T.; Snoek, C. G. Empowering Visual Categorization with the GPU. *IEEE Trans. Multimedia* **2011,** *13* (1), 60–70.
82. Wang, J.; Li, Y.; Zhang, Y.; Xie, H.; Wang, C. *Bag-of-features Based Classification of Breast Parenchymal Tissue in the Mammogram Via Jointly Selecting and Weighting Visual Words.* Paper presented at the Image and Graphics (ICIG), 2011 Sixth International Conference on, 2011.
83. Wang, S.; McKenna, M.; Wei, Z.; Liu, J.; Liu, P.; Summers, R. M. Visual Phrase Learning and Its Application in Computed Tomographic Colonography. *Medical Image Computing and Computer-Assisted Intervention–MICCAI 2013*; Springer, 2013; 243–250.
84. Wang, W.; Liang, D.; Chen, Q.; Iwamoto, Y.; Han, X.-H.; Zhang, Q.; . . . Chen, Y.-W. Medical Image Classification Using Deep Learning. *Deep Learning in Healthcare*; Springer, 2020; pp 33–51.

85. Xiao, J.; Ehinger, K. A.; Hays, J.; Torralba, A.; Oliva, A. Sun Database: Exploring a Large Collection of Scene Categories. *Int. J. Comput. Vision* **2014,** 1–20.
86. Yanagihara, R. T.; Lee, C. S.; Ting, D. S. W.; Lee, A. Y. Methodological Challenges of Deep Learning in Optical Coherence Tomography for Retinal Diseases: A Review. *Transl. Vision Sci. Technol.* **2020,** *9* (2), 11–11.
87. Yang, W.; Lu, Z.; Yu, M.; Huang, M.; Feng, Q.; Chen, W. Content-based Retrieval of Focal Liver Lesions Using Bag-of-visual-words Representations of Single- and Multiphase Contrast-enhanced CT Images. *J. Digit Imag.* **2012,** *25* (6), 708–719. doi:10.1007/s10278-012-9495-1
88. Zhang, J.; Xie, Y.; Wu, Q.; Xia, Y. Medical Image Classification Using Synergic Deep Learning. *Med. Image Analy.* **2019,** *54,* 10–19.
89. Zhang, Z.; Cao, C.; Zhang, R.; Zou, J. *Video Copy Detection Based on Speeded Up Robust Features and Locality Sensitive Hashing.* Paper presented at the Automation and Logistics (ICAL), 2010 IEEE International Conference on, 2010.
90. Zheng, Q.-F.; Wang, W.-Q.; Gao, W. *Effective and Efficient Object-based Image Retrieval Using Visual Phrases.* Paper presented at the Proceedings of the 14th annual ACM international conference on Multimedia, 2006.
91. Zhu, L.; Rao, A.; Zhang, A. Advanced Feature Extraction for Keyblock-based Image Retrieval. *Info. Syst.* **2002,** *27* (8), 537–557.
92. Zhu, L.; Zhang, A.; Rao, A.; Srihari, R. *Keyblock: An Approach for Content-based Image Retrieval.* Paper presented at the Proceedings of the eighth ACM international conference on Multimedia, 2000.

CHAPTER 6

A Survey on Evolutionary Algorithms for Medical Brain Images

NURŞAH DINCER[1] and ZEYNEP ORMAN[2]

[1]*Department of Computer Programming, School of Advanced Vocational Studies, Dogus University, 34680 Istanbul, Turkey*

[2]*Department of Computer Engineering, Istanbul University-Cerrahpasa, 34320 Avcilar, Istanbul, Turkey*

*Corresponding author. E-mail: ormanz@istanbul.edu.tr

ABSTRACT

Evolutionary algorithms (EAs) are the subject of artificial intelligence studies in computer science. These algorithms which simulate the change in nature are applied to traditional computer algorithms. Some of the EAs that adopt this idea are genetic algorithms (GAs), bee colony algorithm, Firefly algorithm (FA), particle swarm optimization (PSO), bacteria foraging algorithm, etc. EAs are used in many fields such as computer networks, image processing, artificial intelligence, cluster analysis. In this chapter, the studies that are conducted between 2014 and 2019 on the application of EAs to 2D MR images are examined according to in which stage of image processing these algorithms are used, the publication year of the articles, and classification of accuracy rates.

6.1 INTRODUCTION

Brain MR images are one of the most commonly used image types in the field of biomedical image processing. Today, many diseases such as cancer, schizophrenia (SZ) can be diagnosed by scientists on these images.

However, the duration of these manual diagnoses and the accuracy of the diagnosis may vary depending on the person's experience. Therefore, computer-aided studies are needed in this field and there are many papers in the literature which have been studied on this subject.[19,38,40]

Arunachalam and Savarimuthu proposed a computer-aided brain tumor detection and segmentation method.[7] The proposed system has stages of enhancement, conversion, feature extraction, and classification. Brain images are enhanced using shift-invariant shearlet transformation (SIST). Brain tumor detection is a difficult task because the brain images contain large variations in shape and density. Shanmuga Priya and Valarmathi focused on edema and tumor segmentation based on skull extraction and kernel-based fuzzy c-means (FCM) approach.[49] The clustering process was developed by combining spatial information-based multiple kernels. Sajid et al. presented a deep learning-based method for brain tumor segmentation using different MRIs.[47] The proposed convolutional neural network (CNN) architecture uses a patch-based approach and takes local and contextual information into account when estimating the output tag. Patil and Hamde proposed a computer-aided system based on monogenic signal analysis for the recognition of brain tumor image.[42] Textural identifiers from different monogenic components were obtained using a completed local binary pattern and a gray-level co-occurrence matrix. Kebir et al. presented a complete and fully automated MRI brain tumor detection and segmentation methodology using the Gaussian mixture model, FCM, active contour, wavelet transform, and entropy segmentation methods.[23] The proposed algorithm consists of skull extraction, tumor segmentation, and detection sections.

As can be seen from the above studies, brain images are difficult to be studied due to their anatomy. Therefore, it has been found more useful to use many methods together. EAs are also frequently used as alternative methods in many studies with brain images. In this chapter, studies using EAs on 2-D brain MR images are presented and various results are discussed.

6.2 BACKGROUND

6.2.1 EVOLUTIONARY ALGORITHMS AND APPLICATIONS

The motivation behind the EAs is to computationally utilize biological evolution, including mainly the natural mechanisms of life, mutation, selection, and more to solve living problems.[58]

EAs are metaheuristic optimization algorithms that work on the concept of population. Metaheuristics are a low-level procedure that can perform a partial search or high-level procedures aimed at finding, producing, or selecting intuitions. They can be applied to a variety of optimization problems with limited processing capability and insufficient or incomplete information. In such cases, they offer a good enough solution.[54] EAs are part of a more comprehensive set of algorithms (EC-evolutionary computation) and are based on random searches and meta-intuition.[58]

More iterations are often required for the accuracy of optimized candidate solutions obtained with EAs. However, there is no guarantee that more iterations will always reduce the error.[58]

Following are the basic steps of EAs to find the best solution for each iteration:

Step 1: After natural selection, the fitness of the population of individuals grows with the effect of environmental pressures.

Step 2: Each individual is evaluated by using the fitness function given by the problem.

Step 3: Parents are selected according to their fitness values between the individuals.

Step 4: New individuals are also produced from the parents by recombination (Step 3).

Step 5: In the selection of future generations, the comparison of eligibility values of old candidates and new individuals is used.

Step 6: If the solution error in all operations is greater than expected, return to Step 1 and the iteration is terminated.[58]

All EAs work on the common principle of the simulated evolution of the individual using selection, mutation, and reproduction processes. However, the algorithms may be different depending on the application and the forms in which they are used.[54]

Some of the best-known types of EAs are differential evolution (DE), differential search algorithm (DSA), genetic programming (GP), evolutionary programming (EP), evolution strategy (ES), genetic algorithm (GA), gene expressing programming (GEP). In addition to the mentioned EAs, there are also swarm intelligence algorithms that consider animals as swarm samples. Some of these algorithms are the ant colony algorithm, bee colony algorithm, cuckoo bird algorithm, particle swarm optimization (PSO).[58]

There are many areas in which the EAs are used in the literature. One of these areas is the brain MRI processing. In the following section, the studies between 2014 and 2019 using 2-D brain MR images and EAs are mentioned.

6.3 BRAIN IMAGE APPLICATIONS AND STUDIES OF EVOLUTIONARY ALGORITHMS

The use of EAs in biomedical image processing studies helps to achieve more successful results. Panda et al. used gray gradient information on brain MR images for thresholding (2016). Because there are many regions in the brain images, thresholding was performed at multiple levels. In the proposed method, a 2-dimensional histogram-based gray gradient is calculated and thus more edge information is preserved. Gray gradient information between pixel values and pixel mean values is used to minimize the loss of information. An evolutionary computational technique was used to optimize gray gradient information to determine optimal multilevel threshold values. For this purpose, a new adaptive swallow swarm optimization (ASSO) algorithm has been applied to the images. The performance of ASSO was found to be better than swallow swarm optimization (SSO).

Narayanan et al.[36] performed a study on brain tumors (2019). They have developed a new algorithm that uses two optimization techniques: PSO and bacterial foraging optimization (BFO) to clearly identify tumor regions and to segment tissues. Contrast limited adaptive histogram equalization (CLAHE) was used for preprocessing of brain MR images, and clustering of the contrast-enhanced image was performed with the modified fuzzy c-means (MFCM) algorithm. The local best and the global best positions in the clustered image are defined by the PSO algorithm, the local best parameter helped BFOA to find the best location values from which the search would be initiated by BFOA. MFCM used the threshold value of BFOA and the best global value of PSO to reassess the clustering result. As a result of these procedures, tissue structures were determined and the tumor area was determined using the proposed algorithm PSBFO-MFCM. In order to prove the power of the algorithm, evaluation parameters of the most advanced techniques were compared and the algorithm has obtained more successful results than other methods.[24,13,45]

Sarkar et al. applied a new unsupervised segmentation method on natural images and medical images, including brain MR images, to improve the distinction between objects within the framework of multi-objective optimization.[48] A multi-objective evolutionary algorithm (MOEA) based on image segmentation technique using multilevel minimum cross-entropy and Rényi entropy has been proposed. MOEA/D-DE (decomposition-based MOEA with DE), one of the MOEAs, is used to determine the optimal solution set instead of the existing single-targeted optimization techniques. The thresholds used in multilevel segmentations were obtained from approximate Pareto Fronts (PF) produced with MOEA/D-DE. The performance of MOEA/D-DE has been compared to single-target and multipurpose optimizers that are inspired by other popular nature. The performance of the proposed algorithm was also tested on MR images containing brain tumors.[12,14]

Vishnuvarthanan et al. have developed a method that uses both optimization and clustering techniques to identify tumor areas on brain MR images.[56] In this study, a new modified fuzzy k-means (MFKM) algorithm based on Bacteria Foraging Optimization (BFO) is proposed. It has been seen that the MFKM algorithm together with BFO improves segmentation on brain MR images. Compared to PSO-based FCM and FCM techniques, the BFOA-based MFKM method has greatly reduced computational complexity. The proposed methodology was evaluated by using comparison parameters.

Agrawal et al. proposed a new method for intracranial segmentation with optimum boundary point detection (OBPD) using pixel density values of brain MR images.[1] First, the skull part of the brain was removed from the images. Two border points were needed to divide the brain pixels into three regions according to their density. (Three regions mentioned: gray matter (GM), white matter (WM) and cerebrospinal fluid (CSF)). The proposed GA-BFO hybrid algorithm was used to calculate the final cluster centers of the FCM method, and thus optimal boundary points were obtained. Other soft calculation techniques GA, PSO, and BFO were also used for comparison.

Vishnuvarthanan et al. used BFO and MFCM algorithms in this study.[57] BFO was used for optimization and MFCM, the advanced version of the FCM algorithm, was used for clustering. Both techniques are well used in a single frame for MRI image segmentation, thus, effective tumor detection and tissue segmentation were obtained simultaneously. Frequently, the parameter setting is not required in the proposed algorithm combination. Therefore, since it increases both manual intervention and high time consumption, it is thought that it will facilitate the work of radiologists in patient diagnostic procedures, with the support provided by an automated algorithm, it is concluded that large volumes of clinical datasets can be easily evaluated.

Chandra and Rao proposed a separate wavelet-based GA to detect tumors in brain MR images.[11] For enhancement, soft thresholding discrete wavelet transform (DWT) and GAs for image segmentation were used. First, MR images were enhanced using a discrete wavelet descriptor. Then, the GA and unsupervised k-means clustering methods were used together to make the most accurate segmentation. A GA was used to determine the best combination of information obtained by the selected criterion. The method was tested on more than 100 real brain MR images. The developed method took advantage of GA's ability to solve optimization problems using a large search area (the label of each pixel of the image).

In this chapter, Oliva et al. proposed a general method for image segmentation.[39] This method consists of minimum cross entropy thresholding (MCET) and crow search algorithm (CSA) methods for image thresholding. In the proposed approach, CSA based on the behavior of crow swarms was used to estimate threshold values. Cross-entropy between classes was minimized by using CSA. CSA encodes a series of candidate threshold points as solutions for each generation. Cross-entropy is used by the objective function to determine the quality of the proposed solution. New candidate solutions are produced using predefined CSA operators in accordance with CSA rules and the value of the objective function. The segmentation quality increases as the process progress. Unlike other optimization techniques used for segmentation recommendations, CSA offers better performance and avoids critical errors such as early convergence to suboptimal solutions and limited exploration–exploitation balance in the search strategy. The proposed method, which is a general segmentation algorithm, provides excellent results in the automatic segmentation of complex MR images. Statistically confirmed experimental results showed that the proposed technique achieved better results in terms of quality and consistency.

Hemanth and Anitha used a modified GA approach to overcome the disadvantage of traditional approaches (2018). These three different GA approaches were applied to the images during the feature selection stage. For all these GA-based methods, the back propagation neural network (BPN) was used as a classifier. Appropriate modifications of existing GA have been made to minimize the randomness of conventional GA. The study focuses on the development of modified reproduction operators that form the core of the algorithm. In this study, different binary processes were used to produce offspring during the crossover and mutation process. Unlike conventional binary operations, these designed binary operations are used in GA for a very random and specific purpose. The application of these approaches was examined in terms of medical image classification. In this study, abnormal brain MR images of four different classes were used. The proposed method provided 98% accuracy compared to other methods.

De et al. performed an application of the GA-based segmentation algorithm to automatically group unlabeled pixels of MR images into different homogeneous clusters.[15] In this method, information about the optimum number of segments before segmentation is not required. With the fuzzy intercluster hostility index, the centroid of the different sections is separated into active/inactive form. The test images are segmented using those selected from these active centroids. Using this method, the optimal number of segments and their respective centroids are obtained. GA method-based fuzzy intercluster hostility index, automatic clustering (ACDE) algorithm using DE and a nonautomatic GA were compared with brain MR images in two different anatomies. The comparison showed that the GA-based automatic image segmentation method is superior to the other two algorithms.

Jothi and Inbarani developed and implemented a supervised hybrid feature selection algorithm called MR tolerance roughset firefly-based quick reduct (TRSFFQR) for MR brain images.[21] With this intelligent hybrid system, it is aimed to take advantage of basic models and to soften its limitations. Different categories of properties, that is, shape, density, and texture-based properties, are obtained from segmented MR images. Hybridization of two techniques, tolerance rough set (TRS) and Firefly algorithm (FA) was used to select the necessary characteristics of a brain tumor. TRSFFQR was compared with Artificial Bee Colony (ABC), Cuckoo Search Algorithm (CSA), supervised tolerance rough set–PSO-based relative reduct (STRSPSO-RR) and supervised tolerance rough set–PSO-based quick reduct (STRSPSO-QR) in terms of performance. As a result, both the efficiency of the technique and its improvements over the currently controlled feature selection algorithms were observed.

Akdemir Akar used bilateral filter (BF) to eliminate Rician noise on MR images as edge protection method in this study.[4] Denoising performance varies according to the selection of BF parameters. For this reason, the parameters of BF were optimized by the GA in the study. The importance of parameter selection in BF was understood by comparing quality parameters such as mean square error (MSE), peak signal to noise ratio (PSNR), signal to noise ratio (SNR), and structural similarity index metric (SSIM) and noise clearing results with other BFs. Experimental results have shown that BF with recommended parameters performs better.

The level set-based Chan and Vese algorithm is a widely used region-based model among active contour models for image segmentation and naturally uses density homogeneity in each region. But, in this model, when the contour is not initialized properly, the possibility of getting trapped in a local minimum is encountered. This problem becomes more critical as said density variations can be found in more varieties and scales on medical images. Mandal et al.[29] proposed a version of the Chan and Vese algorithm independent of the first selection of the contour (2014). In this study, the appropriate energy reduction problem to be solved is formulated using PSO technique, which is one of the metaheuristic optimization algorithms. The algorithm has been successfully applied to both scalar and vector-valued images. Experiments with different types of medical images have shown that the proposed method can significantly improve the quality of segmentation performance obtained by the Chan and Vese algorithm.

As a new method, Huang et al. introduced a new neighborhood intuitionistic fuzzy c-means clustering algorithm with a genetic algorithm (NIFCMGA).[20] This algorithm has the advantages of a heuristical FCM clustering algorithm to maximize benefits and reduce noise/outlier effects. GAs were used to determine the optimal parameters of the algorithm. The proposed technology has been successfully applied to the clustering of different MR and CT image regions that can be expanded to diagnose abnormalities. As a result of the comparisons made with other methods, the performance superiority of the proposed algorithm was revealed.

Ding et al. introduced the multi-agent consensus MapReduce optimization model and coevoluntionary quantum PSO with self-adaptive memeplexes for designing feature reduction method and proposed a multiagent-consensus-MapReduce-based attribute reduction (MCMAR) algorithm.[17] First, coevoluntionary quantum PSO with self-adaptive memeplexes is designed to group particles into different memeplexes aimed at exploring the search area and finding the best region for the reduction of large datasets. Second, the four layers neighborhood radius framework with the compensatory scheme was created to divide large property sets by taking advantage of the interdependence between multifeatured sets. Third, a new multi-agent consensus MapReduce optimization model has been adopted to perform multiple-relevance-attribute reduction using five types of factors for the implementation of the ensemble coevolutionary optimization. Therefore, the uniform mitigation framework of the coevolutionary play of different factors under constrained rationality has been further developed. Fourth, the approximate MapReduce parallelism mechanism was allowed to form the multifactor coevolutionary consensus structure, interaction, and adaptation, which developed different factors to share their solutions. Finally, extensive experimental studies have proven the efficacy and accuracy of MCMAR on some well-known reference datasets. Furthermore, successful applications in large medical datasets are expected to dramatically increase MCMAR in terms of efficiency and feasibility for complex infant brain MRIs.

Mekhmoukh and Mokrani proposed a new method for image segmentation based on PSO and outlier rejection combined with a level set.[32] The proposed algorithm is sensitive to whether the image is noisy or homogeneous and operates based on the initialization of cluster centers. A new FCM version has been developed to improve the outlier rejection and reduce noise sensitivity of the conventional FCM clustering algorithm for image segmentation. In FCM, the first cluster centers are usually randomly selected whereas, in the proposed method, cluster centers were selected optimally with PSO. In addition, spatial neighborhood information is taken into account when performing the calculations. Test procedures of improved kernel possibilistic c-means algorithm (IKPCM) developed in the study were applied with synthetic, simulated, and medical images. This method was compared with different versions of FCM. It has been shown that it provides good segmentation and extraction of various tissues and has improved in terms of its robustness to noise.

Stochastic resonance (SR) is the improvement of low contrast images with noise.[51] In this study, Singh et al. developed a modified neuron model-based SR for brain MR images with T1-weighted, T2-weighted, fluid-attenuated inversion recovery (FLAIR) and diffusion-weighted imaging (DWI) sequences.[51] The multi-objective bat algorithm was used to adjust the parameters of the modified neuron model. The image processing quality varies depending on the selection of these parameters. It was observed that the proposed approach performed well in the improvement of MR images, and as a result, the difference between gray and WM became apparent.

Anaraki et al. proposed a method using CNNs and GA to classify different degrees of gliomas in a noninvasive manner, which is one of the brain tumor types.[6] GAs were used to determine CNN architecture instead of trial and error or to adopt predefined structures. According to the results, the accuracy value of the classification of the three glioma grades is 90.9%. About 94.2% accuracy was obtained in the study in which the tumor types were classified as Glioma, Meningioma, and Pituitary. These results showed that the method is effective in the classification of tumors on brain MR images and because of the flexibility of the method it can help doctors in the early stages of diagnosis.

Manikan et al. performed segmentation of brain MR images using simulated binary crossover (SBX)-based multilevel thresholding with real coded GA.[30] T2-weighted brain MR images were selected for the procedures. The entropy was maximized to achieve optimal multilevel thresholding. The algorithms such as Nelder-Mead simplex, PSO, BF, and ABF were compared with the results of the proposed algorithm. The results showed that the proposed method had better performance for medical images and had a more consistent performance than previous methods.

Kotte et al. applied adaptive wind-driven optimization (AWDO)-based multilevel thresholds for brain MR image segmentation.[26] Images used for image segmentation were selected from axial T2-weighted brain MR images. In this study, the efficacy of AWDO was not investigated for MRI in multilevel thresholds, only a small contribution was made. Optimum multilevel thresholding was achieved by maximizing Kapur's entropy and between-class variance (Otsu's method). In order to investigate the effectiveness of the algorithm, the comparison was made with algorithms such as RGA, GA, Nelder-Mead simplex, PSO, BF, and ABF. The results of the comparisons showed the superiority of the segmentation of the proposed algorithm.

Nayak et al.[37] have proposed a new pathological brain detection system (PBDS) (2018). They used CLAHE to improve the quality of brain MR images. Discrete ripplet-II transform (DR2T) with degree 2 was then applied to the enhanced images. The PCA + LDA approach has been adopted to reduce the large number of coefficients obtained with DR2T. As a final procedure, the MPSO-ELM algorithm obtained from the combination of modified particle swarm optimization (MPSO) and extreme learning machine (ELM) was used for pathological or healthy separation of MR images. The purpose of MPSO in this algorithm is to optimize the parameters of hidden nodes in single-layer layered feedforward networks. The proposed method and other methods were compared with three benchmark datasets. As a result of the comparisons, it has been seen that the proposed method improves the classification accuracy and number of features. Using the MPSO-ELM algorithm, higher accuracy values were obtained than ELM and BPNN classifiers.

Khorram and Yazdi presented an optimized thresholding method that uses the ant colony algorithm for the segmentation of brain MR images.[25] In the algorithm, the textural characteristics of the brain were accepted as heuristic knowledge. The algorithm was designed so that the ants' movements can be more than the nearest eight neighborhoods. In this way, an increase in the ability to discover ants occurred. The applied method showed better performance in post-processing image enhancement based on homogeneity. As a result of the experiments performed with axial T1-weighted brain MR images, a higher accuracy value was obtained compared to conventional heuristic methods, K-means, and expectation maximization.

Pham et al.[41] proposed a new cluster in method for brain MR segmentation (2018). For this purpose, firstly, a new objective function has been found using utilizing kernelized fuzzy entropy clustering with local spatial information and bias correction (KFECSB). Next, an algorithm using an improved PSO with a new fitness function is applied to images for better segmentation. The performance of the proposed method has been tested on a variety of simulated brain MR datasets and real brain MR datasets. As a result of the tests, the method was found to be more effective than the other five states of art methods in the literature. According to the results of the tests, it is seen that it can provide better performance and better results in noisy and inhomogeneous intensity images than other methods.

Ahmed et al. proposed a hybrid method to classify brain MR images as benign and malignant.[3] For this classification process, gray wolf optimizer (GWO) and support vector machine (SVM) with radial basis function (RBF) kernel methods were combined. As a result of the proposed method, the classification accuracy was found to be 98.75%.

Agrawal et al.[2] presented an absolute intensity difference-based (AIDB) technique using adaptive coral reef optimization (ACRO) for brain MR image thresholding (2017). The intensity difference information in the brain image was extracted from the two-dimensional histogram matrix. Since the brain images contain more regions, it is convenient to perform multilevel thresholding. Therefore, the AIDB technique is used for the proposed method. The ACRO method was applied to the images to maximize fitness function. T2-weighted brain MR images were selected from the Harvard medical dataset for the test procedure. According to the results, it is seen that the proposed technique provides better performance than other standard methods.

Nabizadeh et al.[35] proposed an algorithm for segmentation and detection of brain stroke and tumor lesions (2014). For this purpose, they used the histogram-based gravitational optimization algorithm (HGOA). In the algorithm, histogram-based techniques are used to detect initial brain segments. Later, the gravitational optimization-based algorithm was applied to reduce the number of these segments. Finally, thresholding is performed to determine whether it is a tumor or a stroke lesion. In addition, the method is not affected by atlas registration, previous anatomical information or bias corrections because it works independently of these parameters. Accuracy values were 91.5% for the ischemic stroke lesions and 88.1% for the tumor lesions.

In this chapter,[9] proposed an optimized method for processing brain MR images using morphological filters compatible with the human visual system (HVS) (2019). With the logarithmic image processing model, top-line and bottom-line morphological operators were combined for HVS consistency versus filtering. In the morphological filter application, it was necessary to select the structural element with appropriate shape and size in order to detect the tumor correctly. However, this process became difficult as the shape and structure of the tumor may vary according to different stages. Therefore, the structural element was optimized using PSO. Results were evaluated according to parameters such as contrast improvement index (CII), average signal to noise ratio (ASNR), PSNR, and measure of enhancement (EME). According to these results, the evaluation parameter values obtained by enhancement using PSO were higher than those obtained without using PSO.

In this chapter, Virupakshappa and Amarapur performed segmentation and classification of brain MR images using the Modified Level Set approach and Adaptive ANN.[55] For image class estimation, it is crucial to extract useful features from the image. The features extracted for property extraction in the method were multilevel wavelet decomposition features. The classification was performed with adaptive artificial neural network (AANN). Whale optimization algorithm (WOA) was used to optimize ANN. This neural network (AANN) provides optimization of the network structure and provides better classification results for tumors in segmented images. The results obtained using the proposed method were compared with the previous methods. About 98% classification accuracy was obtained with the proposed method.

Subashini et al. (2019) have developed a noninvasive method to identify the degrees of tumors found in brain MR images.[53] Median filter and pulse coupled neural network was used for preprocessing of images. FCM and watershed methods were also applied to the images in the segmentation stage. The tumor was separated from the MR image by Sobels's edge detection and morphological operators. Some extraction techniques have been applied to the images to obtain various features. About 91% classification accuracy was obtained from the system. In addition, this method was found to spend less time in brain tumor grade identification.

Registration, one of the image processing techniques, a lines multiple images and acquires an informative new image.[43] Pradhan and Patra have introduced an original method.[43] The objective of this hybrid method is the optimization of similarity measure in intensity-based nonrigid image registration. However, this method necessitated the optimization of the similarity metric. For this purpose, the bacterial foraging algorithm (BFA) was used to find optimum regional mutual information by the P-spline interpolation method. However, the calculation time for this process was high. Therefore, quantum-behaved particle swarm optimization (QPSO) and BFA have been merged. With this combination, the number of parameters to be optimized was reduced and the calculation time was shortened.

Yang et al.[59] presented a wavelet energy-based method to classify brain MR images as normal or abnormal (2016). Brain images were classified with SVM and weights of SVM were optimized with biogeography-based optimization (BBO). According to the sensitivity and accuracy results, the performance of BBO-KSVM was superior to back propagation neural network (BP-NN), KSVM (kernel SVM), and PSO-KSVM.

Zhang et al.[60] have developed a PBDS for brain MRI (2016). For this purpose, firstly, the extraction of 12 fractional Fourier entropy (FRFE) properties was performed for each of the brain images. The properties were then used in a multilayer perceptron (MLP) classifier. The developments provided by the MLP are as follows: The first determined the optimal number of hidden neurons by the pruning technique. Of these techniques, dynamic pruning (DP), Bayesian detection boundaries (BDB), and Kappa coefficient (KC) were subjected to comparison. Secondly, the adaptive real-coded biogeography-based optimization (ARCBBO) was used for training bias and weights of MLP. The proposed FRFE + KC-MLP + ARCBBO method obtained an average accuracy of 99.53%.

Rajesh et al.[44] presented a system for the detection and classification of brain tumors (2019). Differential based adaptive filtering (DAF) method was used to remove the noise in the images during the preprocessing stage. Skull elimination was also performed using erosion. Segmentation of tumors was utilized region growing algorithm. Rough set theory (RST) has extracted the features of the segmented images. Tumors were also trained and tested with particle swarm optimization neural network (PSONN) to classify them as normal and abnormal. With PSONN, it is aimed to search for training parameters, in other words, decision variables for optimization. PSO was also used with an artificial neural network to minimize MSE and improve the learning process.

Lahmiri[28] compared three systems to detect gliomas on brain MR images (2017). A different PSO technique was used in each of these systems for the segmentation of brain MR images. These were classic PSO, Darwinian particle swarm optimization (DPSO) and fractional-order DPSO (FODPSO). After segmentation, the directional spectral distribution (DSD) signature of these images was calculated. The multi-fractals of the calculated DSD were then obtained by estimation using generalized Hurst exponents. Finally, the classification of these fractal features was performed with SVM. The classification accuracy of these three systems was evaluated using the leave-one-out cross-validation method (LOOM). According to the results, each of the three systems performed better than the previous ones. However, it was considered that the FODPSO-DSD-multi-scale analysis (MSA) system could be a more promising system for the clinical environment because of its high accuracy and low processing time.

Nayak et al.[37] have developed a new PBDS (2018). In this system, CLAHE was used to enhance the quality of brain MR images. The features were extracted with a two-dimensional PCA (2DPCA) method. A compact and discriminative feature set was created using the PCA + LDA combination. The combination of modified differential evolution (MDE) and ELM methods were used to classify images as healthy or pathological. In this combination, input weights and hidden biases of single-hidden-layer feed-forward neural networks (SLFN) have been optimized with MDE. The proposed system was compared to three datasets. According to the results, the proposed method was superior to its equivalents, and the MDE-ELM classifier has better accuracy than conventional algorithms.

Bose and Mali[10] proposed an algorithm that combines FCM and ABC for image segmentation (2016). In this algorithm, the fuzzy membership function is used to find the cluster centers which are optimized by ABC. Compared with other optimization techniques such as PSO, GA, and EM, this new algorithm (FABC) was found to be more efficient. FABC did not depend on the selection of initial cluster centers and provided better performance in terms of convergence, time complexity, robustness, and segmentation accuracy. These have eliminated the disadvantages of FCM. The algorithm has become more efficient by utilizing the random properties of ABC to initialize cluster centers. GA, PSO, EM, and the proposed algorithm were applied to synthetic, tissue and medical images, including brain images. As a result of these applications, the effectiveness of the proposed algorithm has been proved.

Kaur et al.[22] presented a multilevel thresholding method for the automatic segmentation of lesions on brain MR images (2018). In the method, density and edge information found in GLCM and image histograms were used to calculate multiple thresholds. In order to reduce the high computational complexity resulting from search methods, the fitness function had to be optimized. For this purpose, a mutation-based particle swarm optimization (MPSO) technique was used. Also, the search capabilities of this method are better than the conventional version. The performance comparison of the proposed method was performed according to these three different measures. According to the measurement results, the proposed method performed better than the other competing algorithms.

SZ is one of the most important brain diseases worldwide. Most of the analyses are performed according to volumetric measurements on brain MR images. These measurements differ according to the heterogeneity of SZ.[31] Therefore, in this study, the links between schizophrenic MR images and typical images were examined by Manohar and Ganesan.[31] Texture features such as Hu moments, gray level co-occurrence matrix (GLCM), Zernike moments, and structure tensor have been used to represent specific pattern changes in schizophrenic MR images. The distinction between healthy images and schizophrenic images was achieved by using the binary particle swarm optimization (BPSO)-based fuzzy SVM (FSVM) classifier with mutual information quotient as the objective function in the feature selection stage. The skull portion of brain MR images was stripped with a nonparametric region-based active contour. According to the results, it is seen that the proposed method can better separate the brain region from the skull compared to other methods. About 90% accuracy was obtained with BPSO-FSVM. It has a better classification than BPSO-SVM.

Mishra et al.[34] presented a new model for brain tumor detection and classification (2019). An improved fast and robust FCM algorithm has been developed as a segmentation algorithm to smoothen images and to reduce noise in brain MR images. The gray level co-occurrence matrix technique was used to extract features from the images, and these properties were then used in the modified adaptive sine cosine optimization algorithm–particle swarm optimization (ASCA–PSO)-based LLRBFNN model, which was proposed for benign, malignant tumor classification. By optimizing the weights of the LLRBFNN model with the MASCA–PSO algorithm, manual detection of radiologists was avoided. When the comparisons made with different models and classification accuracy values are examined, it was seen that better results are obtained with the proposed model.

6.4 RESEARCH CHALLENGES AND PROSPECTS OF EVOLUTIONARY ALGORITHMS

In this chapter, the studies between 2014 and 2019, in which EAs were applied to 2-dimensional brain MR images, were examined. In order to analyze the effects of these studies on brain MR images, the studies were compared according to the methods used, publication years, datasets and accuracy rates.

EAs have been used in many different stages in the processing of brain MR images. All tables between Table 6.1 and Table 6.7 provide information about which methods are used in which stages. In Table 6.1, the studies using PSO are listed by year and the stage of use. It is examined that PSO is used in most of the image processing stages. In some studies, the PSO algorithm is used in a combination with other EAs. In some other studies, some modified versions of the PSO algorithm were presented.

TABLE 6.1 Brain MR Images Processing which are Used in Studies Using PSO.

Year	Reference	Method(s)	Stage of use
2019	[36]	PSOBFO	Segmentation
2014	[29]	PSO	Image denoising
2018	[17]	PSO	Feature reduction
2015	[32]	PSO	Segmentation
2018	[37]	MPSO	Classification
2018	[41]	PSO	Segmentation
2019	[9]	PSO	Image filtering
2015	[43]	BF QPSO	Image registration
2019	[44]	PSONN	Feature extraction
2017	[28]	classical PSO, DPSO, or FODPSO	Segmentation
2018	[22]	MPSO	Thresholding
2018	[31]	BPSO	Classification
2019	[34]	MASCA–PSO	Feature extraction

In the evaluation of the DE algorithm in Table 6.2, it was found that this algorithm was only used in two studies in 2017 and 2018, respectively. In one of these two studies, the algorithm was used for the segmentation and the other study used it for classification.

TABLE 6.2 Brain MR Images Processing which are Used in Studies Using DE.

Year	Reference	Method(s)	Stage of use
2017	[48]	MOEA/D-DE	Segmentation
2018	[37]	MDE	Classification

The evaluation of the BFO algorithm is given in Table 6.3. Similar to PSO, BFO has been used in conjunction with other EAs in some studies. The stages in which this algorithm is used are segmentation and thresholding.

According to the evaluation in Table 6.4, GA was used in many studies similar to PSO. This algorithm has been utilized in almost every stage of the studies. In related studies, it has been found that GA is used in combination with other methods. In this table, the studies in which the modified versions of GA are used are also given.

TABLE 6.3 Brain MR Images Processing which are Used in Studies Using BFO.

Year	Reference	Method(s)	Stage of use
2017	[56]	BFO	Segmentation
2014	[1]	GA BFO	Segmentation
2018	[57]	BFO	Segmentation
2019	[36]	PSO BFO	Segmentation

TABLE 6.4 Brain MR Images Processing which are Used in Studies Using GA.

Year	Reference	Method(s)	Stage of use
2016	[11]	GA	Segmentation
2018	(Hemanth et al., 2018)	Three GA combinations	Feature selection
2016	[4]	GA	Image denoising
2016	[15]	GA	Segmentation
2015	[20]	NIFCMGA	Segmentation
2019	[6]	GA	Classification
2014	[30]	Real coded genetic algorithm	Segmentation
2014	[1]	GA BFO	Segmentation
2019	[46]	MedGA (Medical image preprocessing based on GAs)	Thresholding
2014	[5]	GA	Preprocessing
2017	[27]	GA	Feature selection
2019	[8]	GA	Feature selection
2019	[33]	GA	Segmentation
2018	[50]	GA	Feature selection
2019	[52]	GA	Feature selection

ACO, another algorithm evaluated in Table 6.5, is also included in various studies. When these studies are examined, it is determined that this algorithm is mostly used in the segmentation and thresholding stages. In addition, it is sometimes used in conjunction with other algorithms and sometimes it is used in a new form with various changes in it.

The BBO algorithm in Table 6.6 was used in two studies. In these studies, this algorithm is included in the classification stage.

Table 6.7 provides information on two separate studies using CSA. The stages in which CSA was used in the studies were obtained as thresholding and image enhancement which are the preprocessing stages.

TABLE 6.5 Brain MR Images Processing which are Used in Studies Using ACO.

Year	Reference	Method(s)	Stage of use
2019	[25]	ACO	Thresholding
2016	[10]	FABC (Fuzzy-based artificial bee colony optimization)	Segmentation

TABLE 6.6 Brain MR Images Processing which are Used in Studies Using BBO.

Year	Reference	Method(s)	Stage of use
2016	[59]	BBO	Classification
2016	[60]	ARCBBO	Classification

TABLE 6.7 Brain MR Images Processing which are Used in Studies Using CSA.

Year	Reference	Method(s)	Stage of use
2017	[39]	CSA	Thresholding
2017	[18]	CSA	Image enhancement

The EAs given in Table 6.8 illustrate the stage of use of these algorithms in the related studies. According to Tables 6.1–6.8, it is obvious that GA and PSO are the most used algorithms in the mentioned years. The algorithms in these tables have often been combined with other methods. In some studies, the methods have been modified and used instead of their original form.

TABLE 6.8 Brain MR Images Processing which are Used in Studies Using Other Evolutionary Algorithms.

Year	Reference	Method(s)	Stage of use
2016	[21]	FA	Feature selection
2017	[51]	Bat optimization (BO)	Image enhancement
2019	[3]	GWO	Classification
2017	[2]	ACRO	Thresholding
2014	[35]	HGOA	Segmentation
2018	[55]	WOA	Classification
2016	[53]	SFLA	Feature extraction
2016	(Panda et al., 2016)	ASSO	Thresholding
2019	[16]	Social group optimization (SGO)	Thresholding
2018	[26]	AWDO	Segmentation

Table 6.9 shows the comparison of the studies by publication years. As a result of this comparison, it is observed that studies including EAs that are performed on brain MR images are mostly published in 2016, 2018, and 2019.

TABLE 6.9 Comparison of Studies by Years.

Year	Publishing	Total
2014	[1], [29], [30], [35], [5]	5
2015	[20], [32], [43]	3
2016	(Panda et al., 2016), [11], [15], [21], [4], [53], [59], [60], [10]	9
2017	[48], [56], [39], [51], [2], [28], [18], [27]	8
2018	[57], (Hemanth et al., 2018), [17], [26], [37], [41], [55], [37], [22], [31], [50]	11
2019	[6], [25], [3], [44], [34], [16], [46], [8], [33], [52], [9]	11

In Table 6.10, studies are compared by datasets. Harvard Medical University, BrainWeb Simulated, and several hospital databases are the most preferred.

The accuracy ratio obtained from the studies that were performed on the same dataset in the same image processing stage is also compared. The databases that meet these criteria are the Harvard Medical University and BRATS 2015. The image processing step according to the same criteria isdetermined as classification.

In Table 6.11, datasets, classification methods, and accuracy rates of the studies are given.

6.5 FUTURE RESEARCH DIRECTIONS

According to the results obtained from the studies, it can be mentioned that EAs have an important impact in this field. These algorithms have helped other methods used together in these studies and provided better results. There are many examples where it is sometimes used in conjunction with other methods or other EAs. In addition, it is observed that these algorithms are used by making changes in their original states. In this way, better results could be obtained from studies with hybrid or modified models.

TABLE 6.10 Datasets Used in the Related Studies.

Dataset	Publishing
Harvard Medical University	(Panda et al., 2016), [4], [30], [37], [3], [2], [59], [28], [37], [34], [18], [50], [52]
Harvard Brain Web Repository	[36], [56], [57]
BrainWeb Simulated Database	[36], [56], [1], [57], [39], [4], [32], [41]
BRATS	[16]
BRATS 2015	[3], [55], [22]
BRATS 2013	[36], [57]
BRATS 2012	[48], [25], [22]
Internet Brain Segmentation Repository (IBSR)	[1], [32], [41]
Several Hospital Database	(Hemanth et al., 2018), [53], [44], [22], [16], [46], [5], [50], [36]
National Cancer Institute (NCI)	[21]
National Institute of Health (NIH)	[29]
IXI Dataset	[6], [5]
REMBRANDT Dataset	[6]
TCGA-GBM Data Collection	[6]
TCGA-LGG Dataset	[6]
Neuroimaging Tools and Resources (NITRC)	[35]
Whole Brain Atlas (WBA) (WBA 2019)	[9]
IBSR 2019	[9]
Three benchmark datasets, namely, DS-I, DS-II, and DS-III	[37]
National Alliance for Medical Image Computing (NAMIC) database	[31]
ISLES2015	[16]
SICAS Medical Image Repository	[27]

The use of these algorithms on difficult-to-work images, such as brain MRI, has helped improve the results obtained. One reason for the difficulty of working with these data is to create a useful dataset. Sometimes this difficulty can be caused by the fact that the actual datasets are not available or that they do not have the necessary competence to be processed. This has a significant effect on the performance impacts of the proposed methods. Therefore, the researchers are faced with the situation of scanning many sources for a proper dataset and creating the dataset accordingly.

TABLE 6.11 Comparison of Studies Using Other Datasets in Terms of Classification Methods and Accuracy Rates.

Dataset	Publishing	Classification	Accuracy rate
IXI, REMBRANDT, TCGA-GBM, TCGA-LGG	[6]	CNN-GA	94.2%
Medical School of Harvard University (DS-66)	[37]	MPSO-ELM	100%
Medical School of Harvard University (DS-160)	[37]	MPSO-ELM	100%
Medical School of Harvard University (DS-255)	[37]	MPSO-ELM	99.69%
Medical School of Harvard University, BRATS 2015	[3]	GWO-SVM	98.75%
MICCAI, BRATS 2015	[55]	WOA-ANN	98%
Harvard Medical University database	[59]	BBO-KSVM	97.78%
Open access dataset	[60]	BBO-MLP	99.53%
Government Medical College Hospital, Trivandrum, India	[44]	PSONN (Particle Swam Optimization Neural Network)	96%
Medical School of Harvard University (Dataset-160)	[34]	MASCA-PSO	99.875%
Medical School of Harvard University (Dataset-255)	[34]	MASCA-PSO	99.61%

Because of all these difficulties, alternative methods are being considered to create a dataset. Generative adversarial network (GAN) is one of the alternative methods for creating a dataset for this purpose. With GAN, a certain number of real images can be passed through various training phases and original and realistic datasets can be obtained. In this way, a unique dataset can be created both original and without the need for a lot of resources.

6.6 CONCLUSIONS

In this chapter, the studies that use the EAs in the processing of 2D brain MR images are examined. As a result of the studies, it has been found that EAs are utilized in many image processing stages such as segmentation, tumor detection, feature extraction, and classification. When the details of these stages are examined, it is seen that EAs are used in a hybrid way with other methods. In addition to the original versions of these algorithms, various modified versions have been included in the studies. In the mentioned studies, these algorithms are generally used for optimization and improvement of other methods. This optimization has sometimes helped to determine the optimum parameters of the method in which it is used together, sometimes to improve the classification performance and sometimes to obtain more accurate results. As can be seen from the studies examined, EAs have significantly contributed to the studies performed with brain images in the field of biomedical image processing as in other areas.

KEYWORDS

- bioinspired algorithms
- biomedical image processing
- genetic algorithm
- particle swarm optimization
- bacteria foraging optimization
- brain MR segmentation
- brain MR classification

REFERENCES

1. Agrawal, S.; Panda, R.; Dora, L. A Study on Fuzzy Clustering for Magnetic Resonance Brain Image Segmentation Using Soft Computing Approaches. *Appl. Soft Comput.* **2014**, *24*, 522–533.
2. Agrawal, S.; Panda, R.; Samantaray, L.; Abraham, A. A Novel Automated Absolute Intensity Difference Based Technique for Optimal MR Brain Image Thresholding. *J King Saud University Comput. Inf. Sci.* **2017**.
3. Ahmed, H. M.; Youssef, B. A. B.; Elkorany, A. S.; Elsharkawy, Z. F.; Saleeb, A. A.; Abd El-Samie, F. Hybridized Classification Approach for Magnetic Resonance Brain Images Using Gray Wolf Optimizer and Support Vector Machine. *Multimedia Tools Appl.* **2019**, *78*, 27983–28002.
4. Akdemir Akar, S. Determination of Optimal Parameters for Bilateral Filter in Brain MR Image Denoising. *Appl. Soft Comput.* **2016**, *43*, 87–96.

5. Akusta Dagdeviren, Z.; Oguz, K.; Cinsdikici, M. G. Three Techniques for Automatic Extraction of Corpus Callosum in Structural Midsagittal Brain MR Images: Valley Matching, Evolutionary Corpus Callosum Detection and Hybrid method. *Eng. Appl. Artif. Intell.* **2014**, *31*, 101–115.
6. Anaraki, A. K.; Ayati, M.; Kazemi, F. Magnetic Resonance Imaging-Based Brain Tumor Grades Classification and Grading via Convolutional Neural Networks and Genetic Algorithms. *Biocybern. Biomed. Eng.* **2019**, *39*(1), 63–74.
7. Arunachalam, M.; Savarimuthu, S. R. An Efficient and Automatic Glioblastoma Brain Tumor Detection Using Shift-Invariant Shearlet Transform and Neural Networks. *Imaging Syst. Technol.* **2017**, *27*(3), 216–226.
8. Aswathy, S. U.; Devadhas, G. G.; Kumar, S. S. Brain Tumor Detection and Segmentation Using a Wrapper Based Genetic Algorithm for Optimized Feature Set. *Cluster Comput.* **2019**, *22*, 13369–13380.
9. Bhateja, V.; Nigam, M.; Bhadauria, A. S.; Arya, A.; Zhang, E. Y. Human Visual System Based Optimized Mathematical Morphology Approach for Enhancement of Brain MR Images. *J Ambient Intell. Humaniz. Comput.* **2019**.
10. Bose, A.; Mali, K. Fuzzy-Based Artificial Bee Colony Optimization for Gray Image Segmentation. *Signal Image Video Process.* **2016**, *10*(6), 1089–1096.
11. Chandra, G. R.; Rao, Dr. K. R. H. Tumor Detection in Brain using Genetic Algorithm. *Procedia Comput. Sci.* **2016**, *79*, 449–457.
12. Chowdhary, C. L. 3D Object Recognition System Based on Local Shape Descriptors and Depth Data Analysis. *Recent Pat. Comput. Sci.* **2019**, *12*(1), 18–24.
13. Chowdhary, C. L.; Acharjya, D. P. Segmentation and Feature Extraction in Medical Imaging: A Systematic Review. *Procedia Comput. Sci.* **2020**, *167*, 26–36.
14. Das, T. K.; Chowdhary, C. L.; Gao, X. Z. Chest X-Ray Investigation: A Convolutional Neural Network Approach. In *Journal of Biomimetics, Biomaterials and Biomedical Engineering;* Trans Tech Publications Ltd, 2020; Vol. 45, pp 57–70.
15. De, S.; Bhattacharyya, S.; Dutta, P. Automatic Magnetic Resonance Image Segmentation by Fuzzy Intercluster Hostility Index Based Genetic Algorithm: An Application. *Appl. Soft Comput.* **2016**, 47, 669–683.
16. Dey, N.; Rajinikanth, V.; Shi, F.; Tavares, J. M. R.S.; Moraru, L.; Karthik, K. A.; Lin, H.; Kamalanand, K.; Emmanuel, C. Social-Group-Optimization Based Tumor Evaluation Tool for Clinical Brain MRI of Flair/Diffusion-Weighted Modality. *Biocybernet. Biomed. Eng.* **2019**, *39*(3), 843–856.
17. Ding, W.; Lin, C. T.; Chen, S.; Zhang, X.; Hu, B. Multiagent-Consensus-MapReduce-Based Attribute Reduction Using Co-Evolutionary Quantum PSO for Big Data Applications. *Neurocomput.* **2018**, *272*, 136–153.
18. Gong, T.; Fan, T.; Pei, L.; Cai, Z. Magnetic Resonance Imaging-Clonal Selection Algorithm: An Intelligent Adaptive Enhancement of Brain Image with an Improved Immune Algorithm. *Eng. Appl. Artif. Intell.* **2017**, *62*, 405–411.
19. Hemanth, D. J.; Anitha, J. Modified Genetic Algorithm Approaches for Classification of Abnormal Magnetic Resonance Brain Tumor Images. *Appl. Soft Comput. J.* **2019**, *75*, 21–28.
20. Huang, C. W.; Lin, K. P.; Wu, M. C.; Hung, K. C.; Liu, G. S.; Jen, C. H. Intuitionistic Fuzzy *c*-Means Clustering Algorithm with Neighborhood Attraction in Segmenting Medical Image. *Soft Comput.* **2015**, *19*(2), 459–470.

21. Jothi, G.; Inbarani, H. H. Hybrid Tolerance Rough Set–Firefly Based Supervised Feature Selection for MRI Brain Tumor Image Classification. *Appl. Soft Comput.* **2016**, *46*, 639–651.
22. Kaur, T.; Saini, B. S.; Gupta, S. A Joint Intensity and Edge Magnitude-Based Multilevel Thresholding Algorithm for the Automatic Segmentation of Pathological MR Brain Images. *Neural Comput. Appl.* **2018**, *30*(4), 1317–1340.
23. Kebir, S. T.; Mekaoui, S.; Bouhedda, M. A Fully Automatic Methodology for MRI Brain Tumor Detection and Segmentation. *Imaging Sci. J.* **2019**, *67*(1), 42–62.
24. Khare, N., Devan, P., Chowdhary, C. L., Bhattacharya, S., Singh, G., Singh, S., & Yoon, B. SMO-DNN: Spider Monkey Optimization and Deep Neural Network Hybrid Classifier Model for Intrusion Detection. *Electronics* **2020**, *9*(4), 692.
25. Khorram, B.; Yazdi, M. A New Optimized Thresholding Method Using Ant Colony Algorithm for MR Brain Image Segmentation. *J Digital Imaging.* **2019**, *32*(1), 162–174.
26. Kotte, S.; Pullakura, R. K.; Injeti, S. K. Optimal Multilevel Thresholding Selection for Brain MRI Image Segmentation Based on Adaptive Wind Driven Optimization. *Measurement* **2018**, *130*, 340–361.
27. Kumar, S.; Dabas, C.; Godara, S. Classification of Brain MRI Tumor Images: A Hybrid Approach. *Procedia Comput. Sci.* **2017**, *122*, 510–517.
28. Lahmiri, S. Glioma Detection Based on Multi-Fractal Features of Segmented Brain MRI by Particle Swarm Optimization Techniques. *Biomed. Signal Process. Control.* **2017**, *31*, 148–155.
29. Mandal, A. D.; Chatterjee, A.; Maitra, M. Robust Medical Image Segmentation Using Particle Swarm Optimization Aided Level Set Based Global Fitting Energy Active Contour Approach. *Eng. Appl. Artif. Intell.* **2014**, *35*, 199–214.
30. Manikandan, S.; Ramar, K.; Iruthayarajan, M. W.; Srinivasagan, K. G. Multilevel Thresholding for Segmentation of Medical Brain Images Using Real Coded Genetic Algorithm. *Measurement* **2014**, *47*, 558–568.
31. Manohar, L.; Ganesan, K. Diagnosis of Schizophrenia Disorder in MR Brain Images Using Multi-objective BPSO Based Feature Selection with Fuzzy SVM. *J Med. Biol. Eng.* **2018**, *38*(6), 917–932.
32. Mekhmoukh, A.; Mokrani, K. Improved Fuzzy C-Means Based Particle Swarm Optimization (PSO) Initialization and Outlier Rejection with Level Set Methods for MR Brain Image Segmentation. *Comput. Methods Progr. Biomed.* **2015**, *122*(2), 266–281.
33. Méndez, I. A. R.; Ureña, R.; Herrera-Viedma, E. Fuzzy Clustering Approach for Brain Tumor Tissue Segmentation Inmagnetic Resonance Images. *Soft Comput.* **2019**, *23*(20), 10105–10117.
34. Mishra, S.; Sahu, P.; Senapati, M. R. MASCA–PSO Based LLRBFNN Model and Improved Fast and Robust FCM Algorithm for Detection and Classification of Brain Tumor from MR Image. *Evolut. Intell.* **2019**, *12*(4), 647–663.
35. Nabizadeh, N.; John, N.; Wright, C. Histogram-Based Gravitational Optimization Algorithm on Single MR Modality for Automatic Brain Lesion Detection and Segmentation. *Expert Syst. Appl.* **2014**, *41*(17), 7820–7836.
36. Narayanan, A.; Rajasekaran, M. P.; Zhang, Y.; Govindaraj, V.; Thiyagarajan, A. Multi-Channeled MR Brain Image Segmentation: A Novel Double Optimization Approach Combined with Clustering Technique for Tumor Identification and Tissue Segmentation. *Biocybernet. Biomed. Eng.* **2019**, *39*(2), 350–381.

37. Nayak, D. R.; Dash, R.; Majhi, B. Discrete Ripplet-II Transform and Modified PSO Based Improved Evolutionary Extreme Learning Machine for Pathological Brain Detection. *Neurocomput.* **2018,** *282,* 232–247.
38. Nayak, D. R.; Dash, R.; Majhi, B. An Improved Pathological Brain Detection System Basedon Two-Dimensional PCA and Evolutionary Extreme Learning Machine. *J Med. Syst.* **2018,** *42*(19).
39. Oliva, D.; Hinojosa, S.; Cuevas, E.; Pajares, G.; Avalos, O.; Gálvez, J. Cross Entropy Based Thresholding for Magnetic Resonance Brain Images Using Crow Search Algorithm. *Expert Syst. Appl.* **2017,** *79,* 164–180.
40. Panda, R.; Agrawal, S.; Samantaray, L.; Abraham, A. An Evolutionary Gray Gradient Algorithm for Multilevel Thresholding of Brain MR Images Using Soft Computing Techniques. *Appl. Soft Computi.* **2017,** *50,* 94–108.
41. Pham, T. X.; Siarry, P.; Oulhadj, H. Integrating Fuzzy Entropy Clustering with an Improved PSO for MRI Brain Image Segmentation. *Appl. Soft Comput.* **2018,** *65,* 230–242.
42. Patil, D. O.; Hamde, S. T. Brain MR Imaging Tumor Detection Using Monogenic Signal Analysis-Based Invariant Texture Descriptors. *Arab. J Sci. Eng.* **2019,** *44*(11), 9143–9158.
43. Pradhan, S.; Patra, D. RMI Based Non-Rigid Image Registration Using BF-QPSO Optimization and P-Spline. *AEU Int. J Electron. Commun.* **2015,** *69*(3), 609–621.
44. Rajesh, T.; Suja Mani Malar, R.; Geetha, M. R. Brain Tumor Detection Using Optimization Classification Based on Rough Set Theory. *Cluster Comput.* **2019,** *22*(6), 13853–13859.
45. Reddy, T.; RM, S. P.; Parimala, M.; Chowdhary, C. L.; Hakak, S.; Khan, W. Z. A Deep Neural Networks Based Model for Uninterrupted Marine Environment Monitoring. *Comput. Commun.* **2020.**
46. Rundo, L.; Tangherloni, A.; Cazzaniga, P.; Nobile, M. S.; Russo, G.; Gilardi, M. C.; Vitabilei, S.; Mauria, G.; Besozzi, D.; Militello, C. A Novel Framework for MR Image Segmentation and Quantification by Using MedGA. *Comput. Methods Progr. Biomed.* **2019,** *176,* 159–172.
47. Sajid, S.; Hussain, S.; Sarwar, A. Brain Tumor Detection and Segmentation in MR Images Using Deep Learning. *Arab. J Sci. Eng.* **2019,** *44*(11), 9249–9261.
48. Sarkar, S.; Das, S.; Chaudhuri, S. S. Multi-Level Thresholding with a Decomposition-Based Multi-Objective Evolutionary Algorithm for Segmenting Natural and Medical Images. *Appl. Soft Comput.* **2017,** *50,*142–157.
49. Shanmuga Priya, S.; Valarmathi, A. Efficient Fuzzy C-Means Based Multilevel Image Segmentation for Brain Tumor Detection in MR Images. *Design Automation Embed. Syst.* **2018,** *22*(1–2), 81–93.
50. Sharif, M.; Tanvir, U.; Munir, E. U.; Khan, M. A.; Yasmin, M. Brain Tumor Segmentation and Classification by Improved Binomial Thresholding and Multi-Features Selection. *J Ambient Intell. Humaniz. Comput.* **2018,** 1–20.
51. Singh, M.; Verma, A.; Sharma, N. Bat Optimization Based Neuron Model of Stochastic Resonance for the Enhancement of MR Images. *Biocybernet. Biomed. Eng.* **2017,** *37*(1), 124–134.

52. Srinivasan, K.; Subramaniam, M.; Bhagavathsingh, B. Optimized Bilevel Classifier for Brain Tumor Type and Grade Discrimination Using Evolutionary Fuzzy Computing. *Turk. J Electr. Eng. Comput. Sci.* **2019,** *27*, 1704–1718.
53. Subashini, M. M.; Sahoo, S. K.; Sunil, V.; Easwaran, S. A Non-Invasive Methodology for the Grade Identification of Astrocytoma Using Image Processing and Artificial Intelligence Techniques. *Expert Syst. Appl.* **2016,** *43*, 186–196.
54. Vikhar, P. A.; Baba, S. S. G. Evolutionary Algorithms: A Critical Review and its Future Prospects, International Conference on Global Trends in Signal Processing, Information Computing and Communication (ICGTSPICC), Jalgaon, India, December 22–24, 2016, IEEE, 2017.
55. Virupakshappa; Amarapur, B. Computer-Aided Diagnosis Applied to MRI Images of Brain Tumor Using Cognition Based Modified Level Set and Optimized ANN Classifier. *Multimed. Tools Appli.* **2018,** *78*(330), 1–29.
56. Vishnuvarthanan, A.; Rajasekaran, M. P.; Govindaraj, V.; Zhang, Y.; Thiyagarajan, A. An Automated Hybrid Approach Using Clustering and Nature Inspired Optimization Technique for Improved Tumor and Tissue Segmentation in Magnetic Resonance Brain Images. *Appl. Soft Comput.* **2017,** *57*, 399–426.
57. Vishnuvarthanan, A.; Rajasekaran, M. P.; Govindaraj, V.; Zhang, Y.; Thiyagarajan, A. Development of a Combinational Framework to Concurrently Perform Tissue Segmentation and Tumor Identification in T1 - W, T2 - W, FLAIR and MPR Type Magnetic Resonance Brain Images. *Expert Syst. Appl.* **2018,** *95*, 280–311.
58. Yar, M. H.; Rahmati, V.; Oskouei, H. R. D. A Survey on Evolutionary Computation: Methods and Their Applications in Engineering. *Modern Appl. Sci.* **2016,** *10*(11).
59. Yang, G.; Zhang, Y.; Yang, J.; Ji, G.; Dong, Z.; Wang, S.; Feng, C.; Wang, Q. Automated Classification of Brain Images Using Wavelet-Energy and Biogeography-Based Optimization. *Multimed. Tools Appl.* **2016,** *75*(23), 15601–15617.
60. Zhang, Y.; Sun, Y.; Phillips, P.; Liu, G.; Zhou, X.; Wang, S. A Multilayer Perceptron Based Smart Pathological Brain Detection System by Fractional Fourier Entropy. *J Med. Syst.* **2016,** *40*(173).

CHAPTER 7

Chatbot Application with Scene Graph in Thai Language

CHANTANA CHANTRAPORNCHAI* and PANIDA KHUPHIRA

Faculty of Engineering, Kasetsart University, Bangkok, Thailand

*Corresponding author. E-mail: fengcnc@ku.ac.th

ABSTRACT

The scene graph is used to represent the semantics of images or visual understanding. It has been used frequently for image retrieval and image generation tasks. We develop a scene graph generator tool from a single image. This tool creates a scene graph representing Thai language. The main methodology contains three steps: image captioning, scene graph parser, and machine translation. We propose an application of chatbot demonstrating the use of the generated scene graph data. The application is developed using Dialogflow. The response is in a JSON form which can be applied further. The image is submitted and the scene graph is generated. Then, the sentence is translated into Thai and by using PyThaiNLP library, the parts of sentences are changed into Thai language. We also show the metric values of the machine translator and caption generator. For the translator model, we use BLEU, GLEU, WER, and TER scores.

7.1 INTRODUCTION

Scene graph proposed by[13] is one type of the graphs, representing the relations between objects inside the image. In the graph, each node represents the objects in the image. The leaf node can be physical, geometric, or material depending on each object type. We can use the scene graph to represent the image semantic. It is useful for image captioning, image generation[12] image retrieval,[23] etc.

In Figure 7.1, an image is shown along with the caption "a black and white dog playing with frisbee." In Figure 7.2, the example of scene graph based on Figure 7.1 is shown. The scene graph contains the details of the objects, relations, and attributes that are extracted from the image and its description. "Dog" and "frisbee" are objects while "plays" is a relation.

Retrieved from:L doglab.com

a black and white dog playing with a frisbee .

FIGURE 7.1 Example image with caption.

A scene graph generator is created from understanding semantic at levels such as sentences or images. Some researchers used more than one semantic level in their work to improve the accuracy of the generator. Considering the sentence levels, most of the current tasks are based on English language. Localizing the graph in other languages, such as Thai language has not been found yet. In this chapter, we demonstrate the development of the scene graph generator that can be applied to the Thai language. The derived scene graph data set can be alternatives for Thai developers to create tasks such as Thai image captioning applications. The method contains the following steps, given an image as an input of the generator. The input image is given to the caption generator and the output sentence is fed to the scene graph parser to put the information in the scene graph format. Finally, the scene graph in English language is translated to Thai language by a neural machine translator.

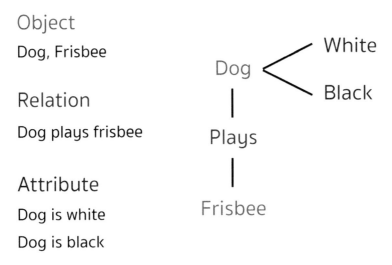

FIGURE 7.2 Scene graph example.

7.2 BACKGROUND

The scene graph is the graph structure which describes relations or attributes between two objects.[17] There are various ways to develop the scene graph generator.[26] The first approach is to generate captions using convolutional neural network (CNN) and recurrent neural network (RNN).[14,20] Then, the caption is used to be the input of the graph parser to convert into the scene graph (). The second approach is to use object detection and use attribute extraction as well as relation extraction to generate the scene graph. The generator utilizes the object detection scheme and then uses feature extraction to convert the information into the scene graph.[3]

For two basic tasks such as object detection, and object recognition, COCO data set is one of popular data sets utilized.[25] CNN is the popular network used in developing image recognition tasks including face recognition,[1] land use classification,[32] object recognition,[11] etc. Various kinds of CNNs include AlexNet,[16] GoogLeNet,[24] ResNet,[9] etc. For the object recognition task, the known network is RCNN whose performance was improved to be Fast-RCNN[8] and Faster-RCNN.[22] There are many researches improving the performance of object detection models such as single shot detector (SSD)[19] likes Faster-CNN and YOLO.[21]

Li et al. presented Factorizable Net which considers subgraph-based. Subgraphs are generated from the fully-connected graph where the edges can refer to similar phrase regions.[17] In their previous work,[18] they generated the scene graph from all objects and relations from an image by using their novel neural network model which is called MSDN model. Yang et. al.[30] created a graph RCNN which consists of three steps: object node extraction, relationship edge pruning, and graph context integration. These works[18,30] considered visual gnome as data set for training.[15]

Xu et.al. applied RNN and use iterative message passing to improve the scene graph prediction. They predicted both objects and relationships based on visual gnome data set (Xu et.al, 2017). The model takes an image as an input, then generate RPN proposals which are passed through their inference model, RNN. The RNN contains GRU cells for nodes and edges connected to each other. The messages are passed through these nodes, pooled and sent to the next nodes. Ref [29] presented Relationship Context-InterSection Region (CISC). They focused on the intersection region of object bounding boxes for feature extraction. The intersection may imply the interactive parts among objects.[5] The approach is based on RNN which utilizes memory and message passing.

Since scene graph generation requires lots of training data, especially relationships between objects in images, large effort is needed to label relationships. The relationship labels in the data set are usually missing. Chen et.al. (Chen et al., 2019) proposed a generative model to predict scene graph labels with limited labels in images. The approach is based on semi-supervised learning. The approach can be used to create scene graph labels and predicate classification.

Table 7.1 represents the different inputs and outputs of these previous works which create the scene graph. Currently, the scene graph data set supported only in English language such as Visual Gnome. This data set contains a lot of object information on an image like a coordinate of each object. However, the data set in the scene graph format still has no support in other languages. In our chapter, we utilize the existing scene graph generator and augment the steps for translation into a local language.

TABLE 7.1 Previous Scene Graph Generation Approach.

Model	Input	Output
F-Net [17]	Image and RPN proposal	Object and relation
MSDN [18]	Image, object proposal and Region proposal	Caption and scene graph
RePN, aGCN [30]	Image	Attentional graph, convolution network
Iterative message passing (Xu et al., 2017)	Image	Scene graph

7.3 METHODOLOGY

Figure 7.3 describes the overall steps of this research. The scene graph generator is made up of three elements: caption generator, scene graph parser, and translation machine.[6,4] First, the caption generator model from Show and Tell A Neural Image Caption Generator, which is a public research project of on Github[28] is used. The structure of the caption generator includes image encoder, a deep convolution neural network which is initialized from Inception_v3 checkpoint and hidden layers like Long Short-Term Memory (LSTM). As an initialization for caption generator model, we use image caption generator[27] based on COCO 2014 data set. We use caption 2014 and image 2014 for training and testing data set, evaluation (256 records, 4 and 8, respectively).

Inception_v3 checkpoint is used as a pretrained weight and then the training is done with COCO 2014 dataset. The model is trained with 1,000,000 epochs, which takes around 1 week on our machine with the following specification 8-Core Intel(R) Xeon(R) CPU E5-2680 0 @ 2.70GHz, RAM 126 GB, Harddisk 1 TB, 2 Tesla K40c Memory Usage 12206MiB Power 235W.

The scene graph parser is used[17] to convert the English sentences into the scene graph. The scene graph format includes relations and attributes in JSON format. Stanford Scene Graph Parser is used, the model implemented to support rule-based parser and classifier-based parser. At this point, the machine translator is used. In our case, Py-translate is selected. The library supports python language and is connected to the Google translator API. Structure of Google translate consists of Google's Neural Machine Translation (GNMT) System which is based on LSTM layers.

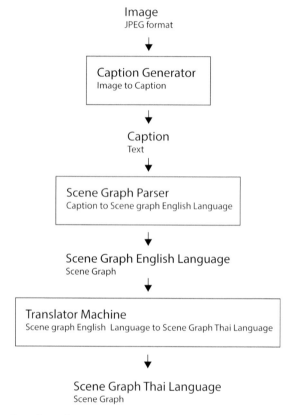

FIGURE 7.3 Overview of preprocess data procedure.

Note that there are two alternatives in applying the translator depending on which may affect the accuracy.[31] For the first approach, the translator machine takes the scene graph which is output from scene graph parser as an input like a word-for-word translation in the top figure of Figure 7.4. In the first step, the sentence is separated into the list of word. Then, each word is put into the translator machine. The last, resulting words are mapped into a result sentence. The second option is to apply the translator from the sentence which is the direct output from the caption generator like sentence-for-sentence translation as in the bottom of Figure 7.4.

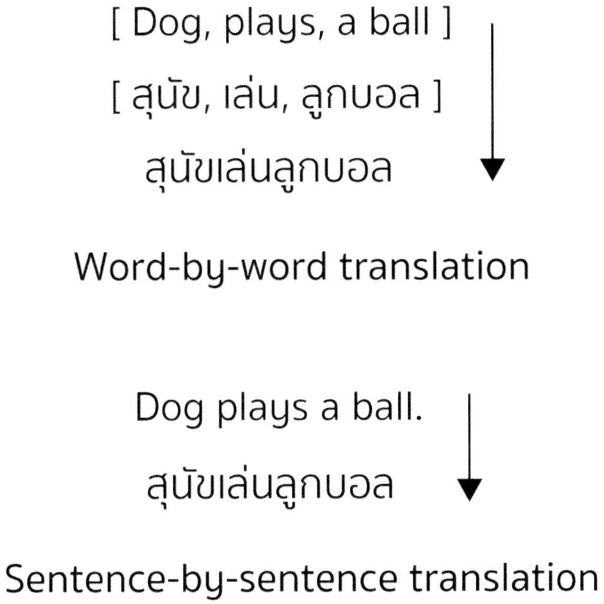

[Dog, plays, a ball]
[สุนัข, เล่น, ลูกบอล]
สุนัขเล่นลูกบอล

Word-by-word translation

Dog plays a ball.
สุนัขเล่นลูกบอล

Sentence-by-sentence translation

FIGURE 7.4 Example of each approach of translation step.

7.4 EVALUATION

Different evaluation metrics are used for each model. Then, we calculate the overall system score by weight sum. For captioning, we use the accuracy based on Microsoft COCO Caption Evaluation module[2] on the evaluated performance of caption generator. COCO val 2014 is used as a testing set which includes 4369 records.

The evaluation modules include metrics: BLEU, METEOR, ROUGE, and CIDEr represents in Table 7.2. For the machine translator, we use NLPMetric.[7] Sub-module's NLPMetric is SPICE, GLEU, WER, and TER.

Bilingual Evaluation Understudy (BLEU) is the measure tool which counts the number of words overlap in resulting translation and compares with the number with the ground-truth translation applied to N-grams. GLEU, also called Google-BLEU, is the minimum of BLEU precision and recall applied to N-gram. Recall is calculated by the number of matching N-grams divided by the number of total N-grams. Word error rate (WER) is used in speech recognition for counting substitutions mainly, calculated from number of error words in the predicted sentence compared with the reference sentence. Translation edit rate (TER) counts the number of edited words which are words deletion, addition, and substitution. The score calculated from the minimum number of edits divided by the average length of reference text.

For Microsoft COCO Caption Evaluation, include BLEU, METEOR, ROUGE-L, CIDEr, and SPICE. Metric for Evaluation of Translation with Explicit Ordering (METEOR) is the harmonic mean of weighted unigram precision and recall which includes stemming and synonym matching.

Recall-Oriented Understudy for Gisting Evaluation (ROUGE), remodels from BLEU adding more attention to recall than precision by paying attention to N-gram. Consensus-based Image Description Evaluation (CIDEr) measures the similarity of resulting sentences against a set of a ground truth sentence by focusing on the sentence similarity by the notions of grammaticality and saliency. Semantic Propositional Image Caption Evaluation (SPICE) is the F1-score of scene graph tuples.

Our experiment includes two options for using a translator machine. Thus, the evaluation module receives an input data set for two ways. The evaluation module gets a predict sentence from a word-for-word translation. The other one is a sentence-for-sentence translation.

The module takes a predicted Thai sentence that was a resulting sentence from translator machine and then, compares with a reference sentence. In addition, we use a Thai language parser or the tokenize tool from PyThaiNLP to separate a predicted sentence into the sentence with a space in between word before feeding to the evaluation model.

The results presented in Table 7.3 is based on TALPCo.[10] TALPCo project was developed based on the main language like Japanese and then this language translated to other Asian languages. The data set translated into English is done by Japanese undergraduate students who had studied at an international junior school and it is rechecked by native British English speaker. The second version of this project supports Thai language. The data set was rechecked correctly by Thai major student at Tokyo University.

Only the first 100 records for evaluating data set are used. The evaluating data set is preprocessed. Our preprocessing step removes the character like a dot from a sentence. Some example of TALPCo data set are "There is a tree in the park.'" which is translated to Thai as "มีต้นไม้อยู่ในสวนสาธารณะ"

From Table 7.2, the highest value for the caption generator is CIDEr which is 0.996. The second is 0.720 from BLEU-4. In translator model, from Table 7.3, the highest evaluate value is WER which is 5.000. The second is 1.1082 from TER average with the second approach.

TABLE 7.2 The Result of Evaluation from Microsoft COCO Caption.

BLEU				METEOR	ROUGE	CIDEr	SPICE
BLEU-1	BLEU-2	BLEU-4	BLEU-4				
0.320	0.419	0.552	0.720	0.258	0.538	0.996	0.183

TABLE 7.3 The Result of Evaluation from NLPMetric.

Model	BLEU avg	GLEU	WER avg	TER avg
Model-1	0.0000	0.0825	5.000	1.1082
Model-2	0.0480	0.2096	4.000	0.8122

The overall system score calculated from two parts giving equal weights. First, we use CIDEr score for a leader of our caption generator score and GLEU score be a leader for our translator machine score. Overall system score for our first approach, a word-for-word translation, is 0.53925, the second, a sentence-for-sentence translation, is 0.6028.

Github of the code and results are available at https://github.com/Bell001/scene-graph-project.git. The code contains implementation example divided into folders:

```
|--CoreNLP
|--application-captures
|--helper
|--measures-model
    |--translator-machine
        |-- NLPMetrics
            |--test
        |--TALPCo
        |--translate-word
|--process-model
|--python-packages
|--test-more
    |--Trans_data_result
```

In the folder "CoreNLP," it contains scene graph parser[17] derived from https://nlp.stanford.edu/software/scenegraph-parser.shtml. We derived the pretrain model and adopted the deployment sample from it.

We develop the python packages in "python-packages" used for Thai language manipulation. It contains the script for Thai word splitting, translating an English sentence to Thai sentence, and for testing the translation.

Folder "helper" contains the source code in java script that make the translation for the scene graph in JSON file obtained from the scene graph parser previously to Thai words using the python package above. We gather the list of relations in Thai words in a dictionary used for the conversion. ThaiNLP is used to Thai word splitting here after the conversion. Then the Thai scene graph is saved in an output JSON file.

Folder "measures-model" contains the code derived from[7] to compute the score of the translator machine. It also keeps the data set from TALPCo in various language including Thai used in the translation code.

The code for the whole process is in folder "process_model." It contains steps presented as a shell scripts in the folder.

1. Process image which takes an input image into the caption generator model[27] based, returns the English caption file and output the English captions in a text file.
2. Take the English caption text file and input to the scene graph parser[17] which generates the scene graph JSON file.
3. Take the scene graph JSON file and translate into Thai scene graph JSON file.

7.5 APPLICATION

The chatbot application is implemented to demonstrate the usage of the scene graph generator. The chatbot developed by using Dialogflow connected with Facebook messenger. Figure 7.5 shows how chatbot works. In this application, the test sentence for the chatbot does not need to be grammatically correct. We focus on the keywords on the test sentence. If the test sentence contains the keyword on the Dialogflow, the word can be changed. The model with mapping image to the scene graph, the size, color channel, format, and color mode of an image affects the result of the caption generator. Currently, our model still supports only JPEG format image.

From Figure 7.5, we test with the sentence: "Tell me the meaning of image" option. The sentence is sent to the webhook server which connects with our model. After that, the webhook replies a scene graph and caption of the image to Dialogflow as a response to the user.

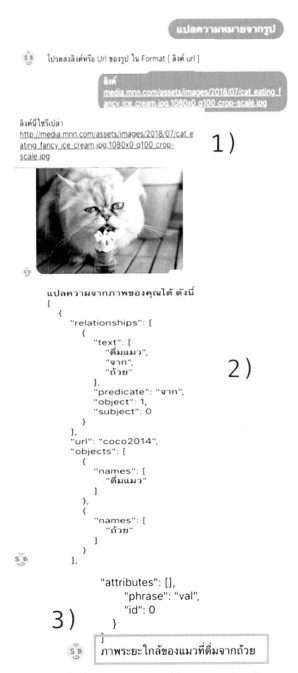

FIGURE 7.5 Messenger Chatbot user response with scene graph and sentences (Example I).

In the Figure 7.6, at (1), the image to create the scene graph is submitted. Then the chatbot replies in JSON format. The responses contain objects and relationship where each word gets translated in Thai. In (3), the Thai sentence is response meaning as "a closed look of a cat drinking from the cup" or "ภาพระยะใกล้ของแมวที่ดื่มจากถ้วย" in Thai.

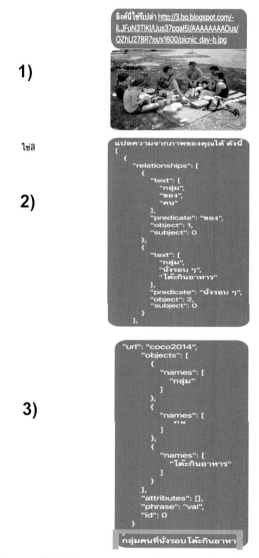

FIGURE 7.6 Messenger Chatbot user response with scene graph and sentences (Example II).

In Figure 7.6, the image submitted at (1) is response with (2) containing three objects and two relationships. The whole sentence is translated as "a group of people sitting around the table" or "กลุ่มคนที่นั่งรอบๆ โต๊ะอาหาร" in Thai.

7.6 CONCLUSION

We present a method for Thai Scene graph generation and the usage on the chatbot. Scene graph contains objects and relations extracted from the given image. The steps contain (1) image captioning (2) scene graph parser (3) Translator machine. The performance is measured for each step and the overall score is computed by the sum of all scores. From our experiment, there are two approaches to use the translator machine. The overall score from a sentence-for-sentence translation gives a higher score than a word-for-word translation. The translator evaluation score implies how correctly the system can translate. The second approach yields better performance as a translator machine.

The results, scene graph in Thai language, show that our scene graph model generation contains a limitation about the accuracy of the scene graph in Thai language. Caption generator is the mainly sub-model that has main impacts on the result. Our model uses the best sentence, which is output from the caption generator, to convert be a scene graph. In our experiment, this model still cannot cover general information due to the limited training data. With this model, the demonstration works on a small group of detected objects and their relations in the image derived by the COCO data set. If the larger data set is available, the approach can be used to generate sentences with larger classes of objects.

KEYWORDS

- **chatbot**
- **scene graph**
- **deep learning**
- **caption generation**

REFERENCES

1. Chantrapornchai, C.; Duangkaew, S. In *Handbook of Research on Deep Learning Innovations and Trends*; A. E. Hassanien et al., Eds.; IGI Global: Hershey, PA, 2019; pp 40–58.
2. Chen, X.; Fang, H.; Lin, T. Y.; Vedantam, R.; Gupta, S.; Dollar, P.; Zitnick, C. L. Microsoft COCO Captions: Data Collection and Evaluation Server, 2015. http://arxiv.org/abs/1709.01507
3. Chowdhary, C. L. Linear Feature Extraction Techniques for Object Recognition: Study of PCA and ICA. *J. Serbian Soc. Comput. Mech.* **2011,** *5*(1), 19–26.
4. Chowdhary, C. L. Intelligent Systems: Advances in Biometric Systems, Soft Computing, Image Processing, and Data Analytics; Apple Academic Press, 2019.
5. Chowdhary, C. L.; Acharjya, D. P. Segmentation and Feature Extraction in Medical Imaging: A Systematic Review. *Procedia Comput. Sci.* **2020,** *167*, 26–36.
6. Chowdhary, C. L.; Goyal, A.; Vasnani, B. K. Experimental Assessment of Beam Search Algorithm for Improvement in Image Caption Generation. *J Appl. Sci. Eng.* **2019,** *22*(4), 691r698.
7. Cunha, G. (n.d.). NLPmetrics. https://github .com/gcunhase/NLPMetrics7
8. Girshick, R. B. Fast R-CNN, 2015. CoRR, abs/1504.08083. http://arxiv.org/abs/1504.08083
9. He, K.; Zhang, X.; Ren, S.; Sun, J. Deep Residual Learning for Image Recognition, 2015. CoRR, abs/1512.03385. http://arxiv.org/abs/1512.03385
10. Hiroki, N.; Okano, K.; Wittayapanyanon, S.; Nomura, J. Interpersonal Meaning Annotation for Asian Language Corpora: The Case of TUFS Asian Language Parallel Corpus (TALPCo). Proceedings of the Twenty-Fifth Annual Meeting of the Association for Natural Language Processing, 2019.
11. Hu, J.; Shen, L.; Sun, G. Squeeze-and-Excitation Networks, 2017. CoRR, abs/1709.01507. http://arxiv.org/abs/1709.01507
12. Johnson, J.; Gupta, A.; Li, F. F. Image Generation from Scene Graphs, 2018. CVPR 2018.
13. Johnson, J.; Krishna, R.; Stark, M.; Li, L.; Shamma, D. A.; Bernstein, M. S.; Fei-Fei, L. Image Retrieval Using Scene Graphs. In 2015 ieee Conference on Computer Vision and Pattern Recognition (cvpr), 2015; pp 3668–3678. DOI: 10.1109/CVPR.2015.7298990
14. Khare, N.; Devan, P.; Chowdhary, C. L.; Bhattacharya, S.; Singh, G.; Singh, S.; Yoon, B. SMO-DNN: Spider Monkey Optimization and Deep Neural Network Hybrid Classifier Model for Intrusion Detection. *Electronics* **2020,** *9*(4), 692.
15. Krishna, R.; Zhu, Y.; Groth, O.; Johnson, J.; Hata, K.; Kravitz, J.; Fei-Fei, L. Visual Genome: Connecting Language and Vision Using Crowdsourced Dense Image Annotations, 2016. https://arxiv.org/abs/ 1602.07332
16. Krizhevsky, A.; Sutskever, I.; Hinton, G. E. ImageNet Classification with Deep Convolutional Neural Networks. In *Advances in Neural Information Processing Systems 25*; Pereira, F., Burges, C. J. C., Bottou, L., Weinberger, K. Q., Eds.; Curran Associates, Inc., 2012; pp 1097–1105.
17. Li, Y.; Ouyang, W.; Zhou, B.; Shi, J.; Zhang, C.; Wang, X. Factorizable Net: An Efficient Subgraph-Based Framework for Scene Graph Generation. ECCV, 2018.

18. Li, Y.; Ouyang, W.; Zhou, B.; Wang, K.; Wang, X. Scene Graph Generation from Objects, Phrases and Region Captions. ICCV 2017, 2017.
19. Liu, W.; Anguelov, D.; Erhan, D.; Szegedy, C.; Reed, S. E.; Fu, C.; Berg, A. C. SSD: Single Shot Multibox Detector, 2015. CoRR, abs/1512.02325. http://arxiv.org/abs/1512.02325
20. Reddy, T.; RM, S. P.; Parimala, M.; Chowdhary, C. L.; Hakak, S.; Khan, W. Z. A Deep Neural Networks Based Model for Uninterrupted Marine Environment Monitoring. *Comput. Commun*. **2020**.
21. Redmon, J.; Divvala, S. K.; Girshick, R. B.; Farhadi, A. You Only Look Once: Unified, Real-Time Object Detection, 2015. CoRR, abs/1506.02640. http://arxiv.org/abs/1506.02640
22. Ren, S.; He, K.; Girshick, R. B.; Sun, J. Faster R-CNN: Towards Real-Time Object Detection with Region Proposal Networks, 2015. CoRR, abs/1506.01497. http://arxiv.org/abs/1506.01497
23. Schuster, S.; Krishna, R.; Chang, A.; Fei-Fei, L.; Manning, C. D. Generating Semantically Precise Scene Graphs from Textual Descriptions for Improved Image Retrieval. Proceedings of the Fourth Workshop on Vision and Language, 2015.
24. Szegedy, C.; Liu, W.; Jia, Y.; Sermanet, P.; Reed, S. E.; Anguelov, D.;. Rabinovich, A. Going Deeper with Convolutions, 2014. CoRR, abs/1409.4842. http://arxiv.org/abs/1409.4842
25. Tsung-Yi, L.; Michael, M.; Serge, B.; James, H.; Pietro, P.; Deva, R.; Lawrence, Z. C. Microsoft Coco: Common Objects in Context. In *Computer Vision – eccv 2014*; Fleet, D., Tomas, P., Bernt, S., Tinne, T., Eds.; Springer International Publishing: Cham, 2014; pp 740–755.
26. Tsutsui, S.; Kumar, M. Scene Graph generation from Images, 2017. http://vision.soic.indiana.edu/b657/sp2016/projects/stsutsui/paper.pdf.
27. Vinyals, O.; et al. Show and Tell: Lessons Learned from the 2015 MS-COCO Image Captioning Challenge. *IEEE Transac. Patt. Anal. Machine Intell*. **2016**, *39*(4), 652–663.
28. Vinyals, O.; Toshev, A.; Bengio, S.; Erhan, D. Show and Tell: A Neural Image Caption Generator, 2014. CoRR, abs/1411.4555. http://arxiv.org/abs/ 1411.4555
29. Wang, Y. S.; Liu, C.; Zeng, X.; Yuille, A. Scene Graph Parsing as Dependency Parsing. NAACL 2018, 2018.
30. Yang, J.; Lu, J.; Lee, S.; Batra, D.; Parikh, D. Graph R-CNN for Scene Graph Generation. ECCV 2018, 2018.
31. Yngve, V. H. Sentence-for-Sentence Translation. Mechanical Translation **1955,** *2*(2), 29–37. http://www.mt-archive.info/MT-1955-Yngve.pdf
32. Zhang, C. Deep Learning for Land Cover and Land Use Classification (Doctoral Dissertation), 2018. DOI: 10.17635/ lancaster/thesis/428

CHAPTER 8

Credit Score Improvisation through Automating the Extraction of Sentiment from Reviews

AADIT VIKAS MALIKAYIL[1], MAHESWARI R.[2*], AZATH H.[3], and SHARMILA P.[4]

[1,2]*VIT Chennai, Chennai, India*

[3]*VIT Bhopal, Bhopal, India*

[4]*Sri Sairam Engineering College, Chennai, India*

[*]*Corresponding author. E-mail: maheswari.r@vit.ac.in*

ABSTRACT

Credit rating firms like D&B, A.M Best Company, etc., usually give scores to companies based on their bank records, scanning the failure to repay the loan, etc. since they only look into the financial details of whether the company defaulted or not in repaying their loans. Text/sentimental analysis improves decisions made by the banks before lending loans to their customers. Also enables businesses to grow profitably by providing information-based intelligence tools. The mission of the work has been to extract the unstructured data from websites (i.e., Glassdoor, Indeed) housing company reviews. The objective is to automate the extraction of the aspects and their corresponding sentiments and cumulate a credit score. This proposed prototype will accept a text input manually or via a text file stacking review. These reviews will be tokenized into words and categorized by noun and adjective. The adjectives are assigned the respective class values/polarity (binary form). The entire goal was to make use of company information stored on the Internet, since it was unaccounted. This kind of information has been extracted from public websites like kanoon.com,

Glassdoor, Indeed.in, etc. So, the rating now is not only based on the bank records but also on how the company operates its employees, sanitation issues, the pay problems if any, beneficial perks given to the employees', etc. Even the consumer whoever given review about the company performance in the market also considered for processing. The sentiments/adjectives given to all noun forms are recorded and given binary score values. The cumulative score of a sentence or a paragraph is then presented in a database. Then a pivot table is generated, which displays a frequency table of the noun forms and their respective sentiment used to describe them. The number of times a noun form has a positive/negative sentiment gets recorded and a score get displayed to the user. Accuracy values for both the text analysis algorithms have been analyzed, and the best one, that is, the TextBlob Analyzer has been put to use since it had accuracy values above 95% for positive sentiments and 91% for the mixed sentiments.

8.1 INTRODUCTION

Businesses make it a good habit of checking the credit score of a company; for instance, when they either invest in the company or purchase shares. The credit score gives enough data to check company compliance with assets and taxes.[1-2] However, this motivates to peruse a proposal that there is so much information on the company data on websites like Indeed.in, Glassdoor, etc. This unstructured data can be used to bring more value to credit rating agencies (CRA's) valuation. It is true that these available unstructured data can be put to use for further analysis. However, websites like Glassdoor do not allow python packages to scrape their pages. Another alternative is using Google Chrome extensions that scrape portions of these pages and save it in a CSV file. But this mere scraping is not efficient for the use case.[3-4] The reason is that the system needs to collect people opinion of a company to analyze of the company's worth qualitatively within a year or two.[5-6] Figure 8.1 shows the sample screen of Glassdoor bans access to scraping.

8.1.1 OBJECTIVE

This work takes into consideration the free, unstructured data and using it for scoring. Sometimes investors do not trust the credit score provided by the CRA.

```
                   return response

E:\softwares\envs\py3\lib\urllib\request.py in      (self, proto, *args)
    568         if http_err:
    569             args = (dict, 'default', 'http_error_default') + orig_args
--> 570             return self._call_chain(*args)
    571
    572 # XXX probably also want an abstract factory that knows when it makes

E:\softwares\envs\py3\lib\urllib\request.py in           (self, chain, kind, meth_name, *args)
    502         for handler in handlers:
    503             func = getattr(handler, meth_name)
--> 504             result = func(*args)
    505             if result is not None:
    506                 return result

E:\softwares\envs\py3\lib\urllib\request.py in              (self, req, fp, code, msg, hdrs)
    648 class HTTPDefaultErrorHandler(BaseHandler):
    649     def http_error_default(self, req, fp, code, msg, hdrs):
--> 650         raise HTTPError(req.full_url, code, msg, hdrs, fp)
    651
    652 class HTTPRedirectHandler(BaseHandler):

HTTPError: HTTP Error 403: Forbidden
```

FIGURE 8.1 Glassdoor bans access to scraping.

One reason is that, companies believe that the score might not be updated to the current date, and that, it could have tampered. Hence, this work looks at getting the most recent data and generating a score that could add value to the general credit score. The proposed system uses python and its Natural Language ToolKit (NLTK) corpus to perform text/sentiment analysis. It made use of data frames from the Pandas package. It has helped to create tables out of lists and dictionaries. This scrutinizes and provides better access to it when the results are computed. The PorterStemmer from the NLTK package is used to stem down words. For instance, "dread," "dreadful," and "dreadfulness" will be considered as word "dread" while computing "dread" as a sentiment. Thus, the mission of the work has been to extract the unstructured data from websites (i.e., Glassdoor, Indeed.in) housing company reviews. The objective is to automate the extraction of the aspects and their corresponding sentiments and cumulate a credit score. The score produced will help to provide value to the credit score generated by the CRAs.

8.2 PROPOSED SYSTEM PLANNING

Firstly, the system needs to take care of the organization of the scraped data. The unstructured data collected needs to be perfectly ordered since the text sentiments describe the particular aspects. Then there comes a need to determine the sentiment, followed by the consideration of interaction factor. Figure 8.2 shows the data flow for the calculation of accuracies from the training text files. The proposed system achieves accuracies from two training text files such as pos_sentiment.txt and mixed_sentiment.txt which has been trained using sentiment analyzer such as Naïve Bayer and Rule-Based.

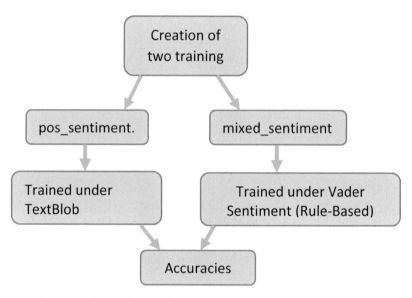

FIGURE 8.2 Calculation of accuracies data flow.

In this system, the user is required to input the number of pages he/she would like to scrape. Once the required pages are set as input, the automation of pages begins, thus the latest reviews will provide valuable insight. Once the scraping is done, the next step is to perform the sentimental analysis on the file containing the scraped data. This system was designed to help aid bring up the value of some CRAs offer. As seen noticeably in some literature surveys, the investors do not entirely trust the CRAs completely.[7-8] The data collected/reflected might be sometimes a couple of years old when the company indeed had a good/bad credit score.[9] Otherwise, the agencies end up giving incomplete information that is not very useful to predict the company future.[10-11] The user can take advantage of the automated scraping to collect and analyze data for practically any company from the many registered in the website indeed.com.[12-20]

8.3 SYSTEM DESIGN

The system used Python 3.6 and other necessary packages. This design is intended to add value to the traditional methods of credit score calculation. With a lot of thought process put into action, the solution makes use of the freely available unstructured data available on websites like Indeed, Glassdoor, etc. The system now helps users to get an insight on the company performance not only quantitatively, but also qualitatively. This work includes the use of two test files in which one file contains full stacked repositories of positive sentiment and the other is stacked with test data for mixed sentiments. These data were collected by automating the scraping process by arranging the text reviews by descending order for the positive reviews and the arrangement for the mixed reviews was obtained by arranging the reviews by ascending order. These two datasets are used to get appropriate accuracies for the two algorithms used, namely TextBlob and Vader Sentiment Analyzer. Proceeding toward the sentiment analysis module, the system has imported the packages required for both the mentioned algorithms.

8.3.1 VADER SENTIMENT ANALYZER

The Vader Sentiment Analyzer package gives output for any given text, mostly in float data type with values ranging from negative "−1" to positive "+1," wherein the parameters are "pos," "neg," "neu," "compound". This work, however, uses the compound value for better estimation. The reason is that a sentence can have a mix of sentiments on one or more aspects. The sample set of compound classification is given in the examples column.

> Examples:
> - I do not like the work experience here, but I am pleased about the salary.
> - The salary is not that good, but the free food menu does cease to surprise me.

So a score for sentences like these does not deserve a highly positive score or a negative score. Hence, the system makes use of a compound score that helps to attain a net score value that would do justice to the output. Further, the graph is used to represent the polarity scores that have been collected over a period of time (as in the Indeed.in website). Adding to that, the proposed system has made sure that the reviews that are scraped are only posted in the current financial year. This was done to make sure the scoring is done in recent basis, since too old data may hamper the results.

8.3.2 TEXTBLOB

Now, consider TextBlob algorithm which works in a way that it produces two kinds of scores, namely the subjectivity score and the polarity score. The subjectivity score helps to identify the number of lines that the text file has to be opinionated and how many of them are related to facts. The polarity score such as positive or negative as the name suggested will give the user an idea of whether the text is positive or negative. Similarly, a graph/plot is made using these polarity values. The diagrammatic representation of score analysis using TextBlob is shown in Figure 8.3.

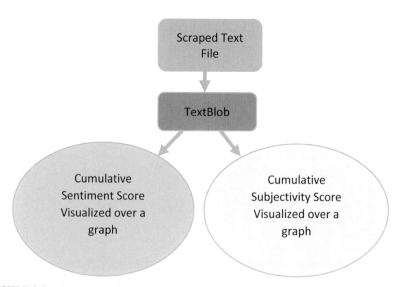

FIGURE 8.3 Score analysis using TextBlob.

8.4 IMPLEMENTATION OF THE POSITIVE SENTIMENT REVIEWS

The training dataset for positive sentiment reviews is considered at first. All these data were scraped from Indeed.in website keeping the ratings descending order as specified by the URL. So accordingly, the soup was created. The screenshot of the created soup is represented in Figure 8.4.

```
soup = make_soup("https://www.indeed.co.in/cmp/Samsung-Electronics-9/reviews?start="+str(page_num)+"&sort=rating_acs")
# for the next button
nxt_btn = soup.find(attrs={"class":"cmp-Pagination-link cmp-Pagination-link--nav"})
```

```
positive_train.txt - Notepad
File  Edit  Format  View  Help
b"Resolved vendor management queries and did trouble shooting with respect to supply chain processes.
Learned how to effectively communicate with stakeholders and establish a rapport with them.
Workplace is fun and there's good work life balance..very good place for work.
friendly environment. worked as a team. everyone is always ready to help each other.
productive work and gives very good pay. overall everything is good..Every day is Day 1 at Amazon.
 You learn something new each day.Very Sophisticated workplace.
Big offices, great cafeteria food, cab benefits, yearly bonuses, good managers and a fun place to consider it as a 2nd home.
 It's a huge family..I would like to tell all about Amazon that its a big brand company.
It is think a lot about their customers as well employers. The atmosphere of the Amazon is friendly.
 Everyone treats to each other like a friend. Love you Amazon & Team..Productive with amiable environment.
 The fun part was interacting with the customers to know more about their day to day problems and some positive feedbacks about amazon ar
It a good place to brush your skills. I had great time working there.
 Its flexible they are strict abt leaves. You need to book leaves in prior if you know the dates.
The Amazon is a good place to work with so talented people and to enhance our knowledge and by working there you will improve your workir
Awesome place to work and learn.The work culture in amazon is great.People are always work driven and love learning and teaching new skil
Life changing experience,Everyday is day zero, you learn everyday,management was helpful & they actually see the positive part in a persc
 The work culture is flexible and open-minded.2. The management takes care of the company needs in an efficient manner.
The number of hours of work do not exceed the daily hours of operation unless you require to work over time.
Lot of benefits in accordance with the amount of work and quality of work we do..Biggest retailer. Best place to work.
Attractive Salary packages in India. Work Culture is great, and high opportunities to grow in the company itself.'
The work environment of the company is very good.People feel proud to be part of the company provide full support to their employees in p
A good place to work over all with a friendly environment. Flexible to work hours though little long.
 Good cafetaria though varities are less but tastly friendly environment , good & well trained employee , also senior are very helpfull,
Amazing place to work.Amazon is an organisation's which treat it's employees very well.
Amazon offers great learning and developing opportunities to its employees and also give opportunities to work onsite (overseas).
Good environment to job in Amazon Good team work.sopting to our team and TL and supervisor .
```

FIGURE 8.4 Created soup for the positive sentiments text file.

Sample reviews with the best rating appearing at first are shown in Figure 8.5. The screenshot of the reviews in ascending order with the least rating reviews showing first. This is the code for the scraping of the reviews for training in ascending order.

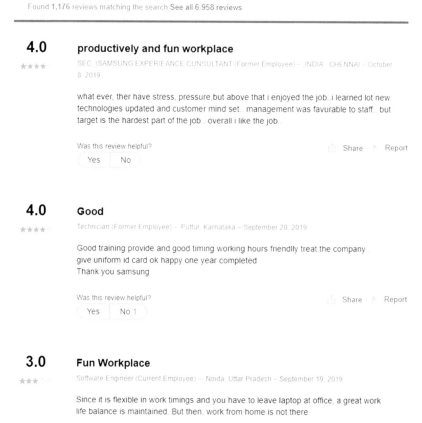

FIGURE 8.5 Positive user reviews.

Next, the system looks at the review scraped in ascending order for the mixed sentiments. The sample screenshot of the reviews is shown in Figure 8.6. Further, the prototype looks at the accuracy obtained by using the Naïve Bayes algorithm with TextBlob. Out of two TextBlob measure parameters such as the polarity and the subjectivity, this system used the polarity to obtain the accuracies. The subjectivity can be used to determine how many fact-oriented or opinion-oriented sentiments are existing in the input file.

8.5 RESULTS AND DISCUSSION

As it observed, the first pair of results are the accuracies for the two approaches which was used to calculate the sentiment scores, namely TextBlob and Vader

Sentiment Analysis. The accuracy attained expending TextBlob module with positive accuracy of 95.95% over 1309 samples and mixed sentiment accuracy of 19.59% via 1118 samples is shown in Figure 8.7.

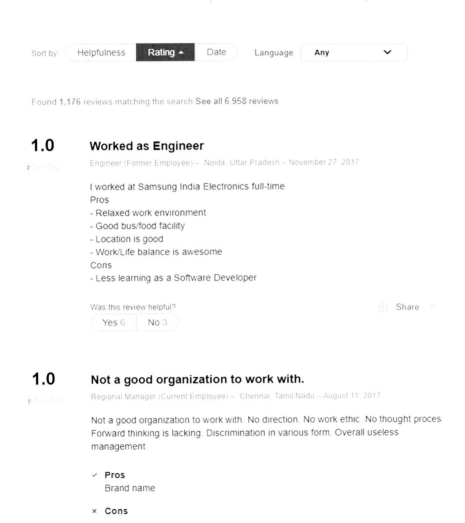

FIGURE 8.6   Mixed user reviews.

Using the Vader Sentiment Analyzer, the system got the following accuracies represented in Figure 8.8. It shows the accuracy achieved paying Vader Sentiment Module with positive accuracy of 89.83% over 1309 samples and mixed sentiment accuracy of 69.85% through 1118 samples. It is inferred that the TextBlob module ensures better accuracy whereas the accuracy is reduced little bit with Vader Sentiment Analyzer as it follows rule-based approach.

```
26  print("Mixed sentiment accuracy = {}% via {} samples".format(neg_correct/neg_

Positive accuracy = 95.9511077158136% via 1309 samples
Mixed sentiment accuracy = 91.59212880143113% via 1118 samples
```

FIGURE 8.7   Accuracy output using TextBlob.

```
27  print( Mixed sentiment accuracy = {}% via {} samples .format(neg_correct/neg_col

Positive accuracy = 89.83957219251337% via 1309 samples
Mixed sentiment accuracy = 69.85688729874776% via 1118 samples
```

FIGURE 8.8   Accuracy output using Vader Sentiment.

As it can be observed, the polarities for the mixed sentiments are a little to the positive side, closer to zero. The reason for this is that as reviews were scraped for training, it was observed that many employees in spite of giving a bad rating for the company also provide good pointers to compensate and to keep their identity safe.

8.5.1   ACCURACIES CALCULATED

The output of the scraped file that was created by automating the scraping process keeping in mind that reviews only of the current year are scraped to ensure data quality is shown in Figure 8.9. Next, the system getting ahead and measures the accuracies for both the approaches used. Making a simple comparison, the system can make out that the accuracy for TextBlob is more than that of Vader.

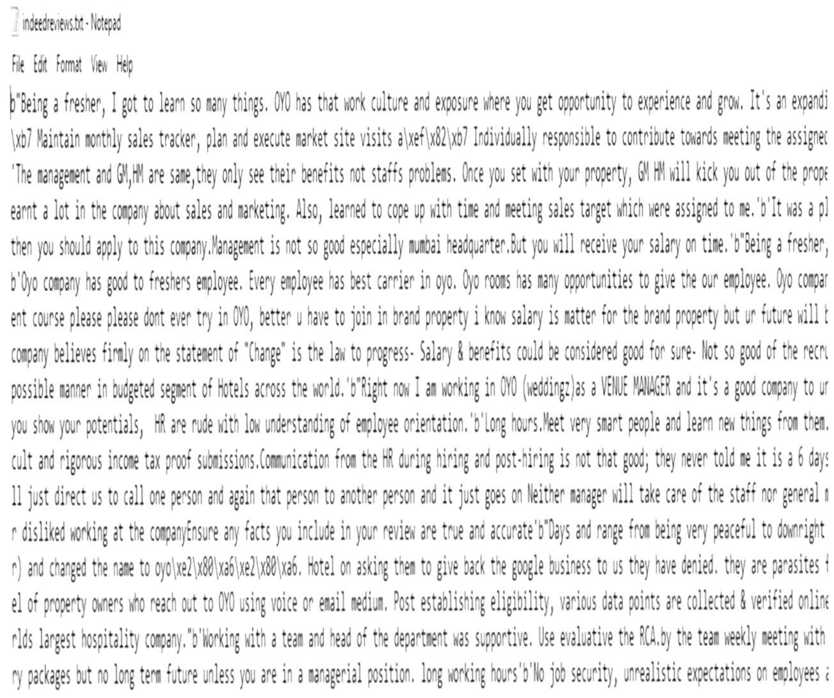
indeedreviews.txt - Notepad

File Edit Format View Help

b"Being a fresher, I got to learn so many things. OYO has that work culture and exposure where you get opportunity to experience and grow. It's an expandi
\xb7 Maintain monthly sales tracker, plan and execute market site visits a\xef\x82\xb7 Individually responsible to contribute towards meeting the assignec
'The management and GM,HM are same,they only see their benefits not staffs problems. Once you set with your property, GM HM will kick you out of the prope
earnt a lot in the company about sales and marketing. Also, learned to cope up with time and meeting sales target which were assigned to me.'b'It was a pl
then you should apply to this company.Management is not so good especially mumbai headquarter.But you will receive your salary on time.'b"Being a fresher,
b'Oyo company has good to freshers employee. Every employee has best carrier in oyo. Oyo rooms has many opportunities to give the our employee. Oyo compar
ent course please please dont ever try in OYO, better u have to join in brand property i know salary is matter for the brand property but ur future will t
company believes firmly on the statement of "Change" is the law to progress- Salary & benefits could be considered good for sure- Not so good of the recru
possible manner in budgeted segment of Hotels across the world.'b"Right now I am working in OYO (weddingz)as a VENUE MANAGER and it's a good company to ur
you show your potentials, HR are rude with low understanding of employee orientation.'b'Long hours.Meet very smart people and learn new things from them.
cult and rigorous income tax proof submissions.Communication from the HR during hiring and post-hiring is not that good; they never told me it is a 6 days
ll just direct us to call one person and again that person to another person and it just goes on Neither manager will take care of the staff nor general n
r disliked working at the companyEnsure any facts you include in your review are true and accurate'b"Days and range from being very peaceful to downright
r) and changed the name to oyo\xe2\x80\xa6\xe2\x80\xa6. Hotel on asking them to give back the google business to us they have denied. they are parasites f
el of property owners who reach out to OYO using voice or email medium. Post establishing eligibility, various data points are collected & verified online
rlds largest hospitality company."b'Working with a team and head of the department was supportive. Use evaluative the RCA.by the team weekly meeting with
ry packages but no long term future unless you are in a managerial position. long working hours'b'No job security, unrealistic expectations on employees a

FIGURE 8.9 Reviews vs scraping.

The reason is that with Vader it uses a rule base implementation for Sentiment Analysis. On the contrary with TextBlob, the system uses Naive Bayes classifier algorithm, which is more efficient. This is the scraped review for performing the sentiment analysis on. The score from these data is used to create graphs and the consumer a complete insight by showing how often the graph reaches the positive and negative peaks.

8.5.2 VADER SENTIMENT VISUALIZATION

The graph produced by the Vader Sentiment Analyzer is shown in Figure 8.10. Notice that the peaks have blunt edges over a range of values. The graph shows peak values at extreme positive and negative polarity values.

With "0" marked as the centre of the y-axis, this makes the output even more apparent.

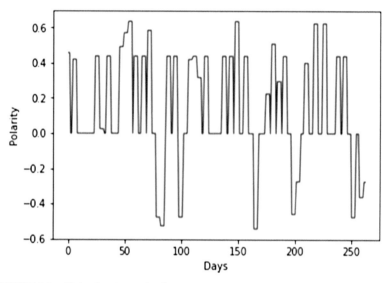

FIGURE 8.10 Vader Sentiment Analyzer.

8.5.3 TEXTBLOB VISUALIZATION

Figure 8.11 represents the graphical outcome produced with the TextBlob package using the Naive Bayes classifier algorithm. It is observable that in the graphical representation, the positive and the negative peaks are pointed, thus giving us more accurate results.

These data were collected keeping in mind every time the year the user is scraping it on. For instance, if the user is scraping it in the year 2019, only those reviews in the particular year will be scraped as the user enters the page numbers in the multiples of 20. Figure 8.12 represents the output achieved through the dates and page number of the reviews scraped.

8.5.4 SCALE OF SUBJECTIVITY OF REVIEWS

The sentiment module reads lines from the text file indeedreviews.txt. Further, it looks at the sentimental analysis module using two algorithms,

Credit Score Improvisation through Automating

one that is rule-based, Vader Sentiment Analyzer and the other that works on top of the Naïve Bayes classifier, called the TextBlob classifier.

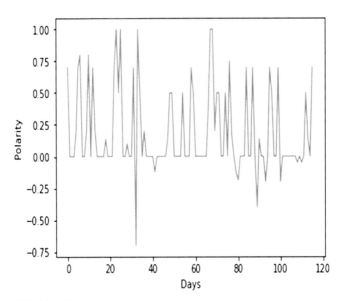

FIGURE 8.11 TextBlob Analyzer.

```
Enter a number in multiples of 20 : 40
this is page 0
October 16, 2019
/cmp/Oyo/reviews?start=20
October 27, 2019
/cmp/Oyo/reviews?start=20
October 17, 2019
/cmp/Oyo/reviews?start=20
October 14, 2019
/cmp/Oyo/reviews?start=20
October 14, 2019
/cmp/Oyo/reviews?start=20
October 10, 2019
/cmp/Oyo/reviews?start=20
October 9, 2019
/cmp/Oyo/reviews?start=20
October 6, 2019
/cmp/Oyo/reviews?start=20
October 4, 2019
```

FIGURE 8.12 Output—Dates and page no. of the reviews scraped.

Figure 8.13 shows an example of the subjectivity of the text. Wherein, the values close to 0.0 are objective and the values close to 1.0 are more subjective.

```
[[0.5, 0.0, 0.0, 0.3, 0.0, 0.3, 0.6000000000000001, 0.6000000000000001, 0.6, 0.0, 1.0, 0.3, 0.06666666666666667, 0.3, 0.0666666
6666666667, 0.0, 0.0, 0.5416666666666666, 0.5416666666666666, 1.0, 1.0, 0.6000000000000001, 0.5, 0.4, 0.0, 0.5, 0.0, 0.0, 0.55,
0.0, 0.0, 0.4, 0.55, 0.45454545454545453, 0.0, 0.0, 0.75, 0.4, 0.625, 0.75, 0.75, 0.75, 0.0, 0.45454545454545453, 0.0, 0.0, 0.
1, 0.0, 0.0, 0.0, 0.0, 0.9, 0.0, 0.125, 1.0, 0.0, 0.0, 0.06666666666666667, 0.0, 1.0, 0.0, 0.375, 0.4, 0.6000000000000001, 0.3,
0.6000000000000001, 0.0, 0.8666666666666667, 0.0, 0.5, 0.45454545454545453, 0.45454545454545453, 0.75, 0.0, 0.9666666666666667,
0.6, 0.75, 0.0, 1.0, 0.65, 0.3, 1.0, 0.3, 0.3, 0.5, 0.4, 0.9, 0.5, 0.0, 0.5, 0.0, 0.0, 0.0, 0.0, 0.0, 0.3, 0.0, 0.0, 0.3, 0.600
0000000000001, 1.0, 0.5, 0.9666666666666667, 0.0, 0.9, 1.0, 0.0, 0.0, 0.0, 0.6000000000000001, 1.0, 0.0, 0.5, 0.0, 0.6000000000
000001, 0.0, 0.0, 0.0, 0.2, 0.6000000000000001, 0.75, 0.0, 0.0, 0.2, 0.75, 0.0, 0.6000000000000001, 0.6, 0.0, 0.1666666666666
66, 0.3333333333333333, 0.3333333333333333, 0.45454545454545453, 0.0, 0.0, 0.0, 0.6000000000000001, 0.3, 0.5, 0.3, 0.0, 0.0, 0.
4, 0.0, 0.0, 0.6000000000000001, 0.6666666666666666, 0.3, 1.0, 0.5, 0.2, 0.0, 0.3, 0.5, 0.0, 0.375, 0.125, 0.0, 0.0, 0.0, 0.0,
0.45454545454545453, 0.5, 0.5, 0.5, 0.0, 0.0, 0.0, 0.5, 0.0, 0.0, 0.0, 0.6000000000000001, 0.5, 0.0, 0.0, 0.0, 0.0, 0.0, 0.6,
0.5, 1.0, 0.3, 0.2, 0.5, 0.5, 0.0, 0.0, 1.0, 0.0, 0.95, 0.3333333333333333, 0.4, 0.375, 0.5, 0.0, 0.0, 0.0, 0.6000000000000001,
0.0, 0.0, 0.6000000000000001, 0.0, 0.6, 0.45454545454545453, 0.0, 0.0, 0.4, 0.0, 0.6000000000000001, 0.8888888888888888, 0.0,
0.0, 0.6000000000000001, 0.1, 0.0, 0.0, 0.25, 0.0, 0.0, 0.0, 0.0, 0.4, 0.0, 0.4, 0.0, 0.5, 0.45454545454545453, 0.0, 0.6000000
```

FIGURE 8.13 Text subjectivity sample.

8.5.5 SCALE OF POLARITY OF REVIEWS

The appropriate analysis is made using the polarity values received at output and shown in Figure 8.14. For instance, the system takes the word "great" for further analysis. As observed, the picture gives probabilistic values for the polarities and subjectivity. The compound values are the ones which are going to make use of in the proposed analysis. Again, the system would probably not use Vander approach since it gives a lesser accuracy as compared to the TextBlob module. Figure 8.15 represents the scale of polarity review sample.

8.5.6 LIMITATION

The proposed system does not have a login module since this was done keeping in mind the fact that the functionality it provides for scraping and text analysis. The system was made to aid and bring value to the credit scores calculated by the CRAs. As of now, the proposed work only assists in scraping all company reviews from Indeed.in website. The limitation of this system is that it is still not capable of scraping through websites,

## Credit Score Improvisation through Automating

namely Glassdoor.com, since the access is forbidden and does not permit python packages to do the same. Secondly, the project does not look at data integrity as a priority. It does not limit the amount of data to scrape, but data integrity is not considered. Large volumes of freely available unstructured data are collected for analysis. Adding permissions to these files would not be a necessity.

word	polarity	subjectivity
great	1.0	1.0
great	1.0	1.0
great	0.4	0.2
great	0.8	0.8

FIGURE 8.14 Polarity vs subjectivity.

```
[[-0.2916666666666667, 0.2, 0.2, 0.0, 0.22727272727272727, 0.0, 0.375, 0.25, 0.2, 0.7, 0.0, 0.2, 0.6, 0.0, 0.7, 0.2, 0.5, 0.0,
0.0, 0.7, 0.0, 0.0, 0.2, 0.0, 0.2, 0.13636363636363635, 0.7, 0.5, 0.5, 0.0, 0.0, 0.0, 0.0, 0.0, 0.2, 0.0, 0.2, 0.7, 0.7, 0.4,
0.0, 0.5, 0.2, -0.16666666666666666, 0.2, -0.16666666666666666, 0.0, 0.0, 0.5, 0.0, -0.2916666666666667, -0.2916666666666667,
0.0, 0.0, 0.7, 0.0, -0.05, 0.0, 0.5, 0.0, 0.2, 0.0, 0.0, 0.0, 0.2, 0.13636363636363635, 0.0, 0.0, 0.8, 0.0, 0.0, 0.8, 0.8
0.0, 0.0, 0.13636363636363635, 0.0, 0.0, 0.0, 0.0, 0.0, 0.0, 0.0, 0.6000000000000001, 0.0, 0.0, 0.0, 0.0, 0.0, -0.16666666666
6666, 0.0, 0.0, 0.9, 0.0, -0.125, 0.0, 0.7, 0.2, 0.7, 0.0, -0.5000000000000001, 0.0, 0.0, 0.13636363636363635, 0.13636363636363635,
0.8, 0.0, -0.2916666666666667, 0.2, 0.2, 0.0, 0.22727272727272727, 0.0, 0.375, 0.7333333333333333, 0.0, 0.8, 0.0, -0.5, 0.35,
0.2, 0.6, 0.2, 0.2, 0.5, 0.1, 0.6000000000000001, 0.4, 0.0, 0.5, 0.0, 0.0, 0.0, 0.0, 0.0, 0.2, 0.0, 0.0, 0.2, 0.7, 1.0, 0.5, 0
7333333333333333, 0.0, 0.6, 1.0, 0.0, 0.0], [0.0, 0.7, 0.0, 0.0, 0.7, 0.0, 0.0, 0.0, 0.2, 0.7, 0.8, 0.0, 0.0, 0.2, 0.8, 0.0, 0
7, 0.2, 0.0, 0.0, 0.0, 0.0, 0.13636363636363635, 0.0, 0.0, 0.0, 0.7, 1.0, 0.5, 1.0, 0.0, 0.0, 0.1, 0.0, 0.0, 0.7, -0.699999999
999998, 1.0, 0.5, 0.0, 0.2, 0.0, 0.0, 0.0, 0.0, -0.125, 0.0, 0.0, 0.0, 0.0, 0.0, 0.13636363636363635, 0.5, 0.5, 0.0, 0.0, 0.0,
0.0, 0.5, 0.0, 0.0, 0.0, 0.7, 0.5, 0.0, 0.0, 0.0, 0.0, 0.0, 0.0, 0.375, 1.0, 1.0, 0.2, 0.5, 0.5, 0.0, 0.0, 0.5, 0.0, 0.75, 0.1
6666666666666, 0.0, -0.125, -0.1875, 0.0, 0.0, 0.0, 0.7, 0.0, 0.0, 0.7, 0.0, -0.4, 0.13636363636363635, 0.0, 0.0, -0.2, 0.0,
0.7, 0.5, 0.0, 0.0, 0.7, -0.2, 0.0, 0.0, 0.0, 0.0, 0.0, 0.0, -0.05, 0.0, -0.05, 0.0, 0.5, 0.13636363636363635, 0.0, 0.7,
0.16666666666666666, 1.0, 0.0, 0.0, -0.2916666666666667, 0.2, 0.2, 0.0, 0.22727272727272727, 0.0, 0.375, 0.2857142857142857, 0
0, 0.7, 0.13636363636363635, 0.2, 0.0, 0.0, 0.2, -0.6999999999999998, 0.0, 0.0, 0.0, 0.0, 0.0, -0.125, 0.0, 0.2], [0.333333333
333333, 0.2, 0.5, 0.0, 0.7, 0.0, 0.0, 0.0, 0.0, -0.3, 0.0, -0.05, 0.0, 0.2, 0.21428571428571427, 0.13636363636363635, 0.0, 0.0
0.8, 0.8, 0.0, 0.2, -0.3, 0.0, 1.0, 0.0, -0.125, 0.0, 0.6, 0.0, 0.0, 0.0, 0.0, 0.7, 0.13636363636363635, 0.0, 0.7, 0.0, 0.0, 0
13636363636363635, 0.0, -0.4, 0.0, 0.0, 0.2, -0.5, 0.0, 0.0, 0.0, 0.7, 0.0, -0.2916666666666667, 0.0, 0.0, 0.0, 0.0, 0.0, 0.7,
0.6, 0.0, 0.0, 0.0, -0.125, 0.2, -0.6999999999999998, 0.0, 0.0, 0.0, 0.05000000000000002, 0.2, 0.0, 0.0, 0.0, 0.0, 0.0, 0.0, 0
0, 0.2, 0.25, 0.0, 0.0, 0.375, 0.2, 0.5, -0.05, 0.0, 0.2, -0.6999999999999998, 0.0, 0.2, 0.0, 0.0, 0.0, 0.13636363636363635, 0
2, 0.0, 0.35, 0.4000000000000001, 0.2, 0.25, 0.0, 0.0, 0.0, 0.5, 0.2857142857142857, 0.0, 0.0, 0.21428571428571427, 0.0, 0.2,
```

FIGURE 8.15 Scale of polarity review sample.

8.6 CONCLUSION AND FUTURE WORK

For inference, it is safe to say that the proposed system will bring some value to the customers looking for legitimate investments. More apt information can be gained from the outputs. Once the peak values from the graphs can be obtained, all values can be recorded. An estimate calculation can be made. Many companies doubt the credit worthiness of an organization given the efficiencies of the CRAs in presenting company compliance to various parameters. A cumulative of many credit reports generated on a company helps generate a company credit report that determines the financial health of an organization. These reports are created to know the credit worthiness of an organization. Accuracy values for both the text analysis algorithms have been analyzed, and the best one, that is, the TextBlob Analyzer has been put to use since it had accuracy values above 95% for positive sentiments and 91% for the mixed sentiments.

KEYWORDS

- **credit rating**
- **sentiment analysis**
- **Glassdoor**
- **TextBlob analysis**
- **cumulative score**

REFERENCES

1. Zhang, M. L.; Peña, J. M.; Robles, V. Feature Selection for Multi-Label Naive Bayes Classification [J]. *Inf. Sci.* **2009**, *179*(19), 3218–3229.
2. Field, B. J. Towards Automatic Indexing: Automatic Assignment of Controlled-Language Indexing and Classification from Free Indexing. *J Doc.* **1975**, *31*, 246–265.
3. Ittner, D. J.; Lewis, D. D.; Ahn, D. D. *Text Categorization of Low Quality Images. Symposium on Document Analysis and Information Retrieval Las Vegas, NV. ISRI*; Univ. of Nevada: Las Vegas, 1995; pp 301–315.
4. Joachims, T. A Probabilistic Analysis of the Rochhio Algorithm with TFIDF for Text Categorization. Machine Learning: Proceedings of the Fourteenth International Conference, 1997; pp 143–151.
5. Lewis, D. D.; Ringuette, M. A Comparison of Two Learning Algorithms for Text Categorization. Third Annual Symposium on Document Analysis and Information Retrieval, 1994; pp 81–93.
6. Ng, H. T.; Goh, W. B.; Low, K. L. Feature Selection, Perceptron Learning, and a Usability Case Study for Text Categorization. Proceedings of the 20th Annual

International ACM SIGIR Conference on Research and Development in Information Retrieval, 1997; pp 67–73.
7. Vijayan, V. K; Bindu, K. R.; Parameswaran, L. A Comprehensive Study of Text Classification Algorithms, Advances in Computing Communications and Informatics (ICACCI) 2017 International Conference, 2017; pp 1109–1113.
8. Yang, Y. Expert Network: Effective and Efficient Learning from Human Decisions in Text Categorization and Retrieval. Proceedings of the 17th Annual International ACM SIGIR Conference on Research and Development in Information Retrieval, 1994; pp 13–22.
9. Yang, Y. An Evaluation of Statistical Approaches to Text Categorization. *Inf. Retr.* **1999**, *1*, 69–90.
10. Maheswari, R.; Sheeba Rani, S.; Sharmila, P.; Rajarao, S. Personalized Secured API for Application Developer. *Adv. Intell. Syst. Comput. Smart Innov. Commun. Comput. Sci.* **2018**, 401–411.
11. Kaur, S.; Kaur, K.; Kaur, P. Analysis of Website Usability Evaluation Methods, International Conference on Computing for Sustainable Global Development, 2016.
12. Gopinath, S.; Senthooran, V.; Lojenaa, N.; Kartheeswaran, T. Usability and Accessibility Analysis of Selected Government Websites in Sri Lanka, 2016.
13. Zhang, H.; Li, D. Naïve Bayes Text Classifier, IEEE International Conference on Granular Computing, 2007.
14. Chowdhary, C. L.; Goyal, A.; Vasnani, B. K. Experimental Assessment of Beam Search Algorithm for Improvement in Image Caption Generation. *J Appl. Sci. Eng.* **2019**, *22*(4), 691–698.
15. Khare, N.; Devan, P.; Chowdhary, C. L.; Bhattacharya, S.; Singh, G.; Singh, S.; Yoon, B. SMO-DNN: Spider Monkey Optimization and Deep Neural Network Hybrid Classifier Model for Intrusion Detection. *Electronics* **2020**, *9*(4), 692.
16. Chowdhary, C. L.; Acharjya, D. P. Segmentation and Feature Extraction in Medical Imaging: A Systematic Review. *Procedia Comput. Sci.* **2020**, *167*, 26–36.
17. Reddy, T.; RM, S. P.; Parimala, M.; Chowdhary, C. L.; Hakak, S.; Khan, W. Z. A Deep Neural Networks Based Model for Uninterrupted Marine Environment Monitoring. *Comput. Commun.* **2020**.
18. Chowdhary, C. L. 3D Object Recognition System Based on Local Shape Descriptors and Depth Data Analysis. *Recent Pat. Comput. Sci.* **2019**, *12*(1), 18–24.
19. Parimala, M.; RM, S. P.; Reddy, M. P. K.; Chowdhary, C. L.; Poluru, R. K.; Khan, S. Spatiotemporal-Based Sentiment Analysis on Tweets for Risk Assessment of Event Using Deep Learning Approach. *J Softw. Pract. Exp.* **2020**.
20. Das, T. K.; Chowdhary, C. L.; Gao, X. Z. Chest X-Ray Investigation: A Convolutional Neural Network Approach. In *Journal of Biomimetics, Biomaterials and Biomedical Engineering*; Trans Tech Publications Ltd, 2020; Vol. 45, pp 57–70.

CHAPTER 9

Vision-Based Lane and Vehicle Detection: A First Step Toward Autonomous Unmanned Vehicle

TAPAN KUMAR DAS

School of Information Technology and Engineering, Vellore Institute of Technology, Vellore, India

*Corresponding author. E-mail: tapan.das@vit.ac.in

ABSTRACT

Automatic driving of a car also proclaimed as the driverless car is perhaps the most fascinating and challenging research of the next decade. Even though the automotive industry has been transformed radically toward automation and remarkable advancement in all the spheres has been realized; however, automatic car driving remains a distant dream in the present era. This task has been put under third-generation artificial intelligence innovation due to the underlying complexity and legal aspects in case of failure. The task of driving on the road is completely a human pursuit; hence, the project involving automation of this task requires the complete spectrum of human ability consisting of all the senses and the motor organs. It involves a series of tasks to be automated. The first of the activity is known as detecting the driving traffic lane automatically. The task becomes challenging due to irregular and inconsistent road conditions. In this chapter, a vision-based sensing and detecting the road are experimented by capturing the front view with the help of a camera mounted on the car. In another embodiment, vehicle detection which is crucial for a driving system to identify is proposed. In this chapter, we have implemented a vehicle identification from the image captured by a camera fitted at the front of the vehicle.

9.1 INTRODUCTION

A system facilitating information about track condition, nearby vehicle position, and on-road pedestrians is considered as an assisting tool en route toward entire or partial automation of driving task. Hence technically, the task is known as lane detection which consists of[2] subtasks: localization of road, identifying vehicle position, and analyzing the position of the vehicle relative to the road. It also includes localizing possible obstacles on the path. We have identified few literatures in this regard; Massimo (2000) discussed the infrastructure need for a vehicle to be intelligent; they are sensors, machine vision, and actuators. Besides it throws light on state-of-the-art technology pertaining "Automatic Road Following." Furthermore, it critically reviews the working of several vision-based systems: SCARF, PVR III, RALPH, ROMA, GOLD, ARGO, and many more. A road consists of multiple driving lanes, and it is highly necessary to mark the pathway on which the vehicle to be rolled. Image processing technique has been found to be quite efficient in this regard[8] which is an integrated approach for dual task to detect the road boundary along with lane marking. Further, lane detection has been addressed by the Caney edge detector.[9] Right and left line detection has been executed by standard Hough transformation within a fixed search area. This works very well for straight as well as slightly curved road.[5-7]

In other hands, vehicle detection technique includes background subtraction,[4] feature-based methods, and frame differentiating by taking low resolution aerial images of cars.[1] Further, vehicle detection from satellite imagery has been studied by Gill and Sharma[3] reveals the total number of vehicles within the desired space of the satellite image by employing morphological image processing, segmentation, and edge detection with an accuracy of 86.5%.

In order to implement the project comprising of lane detection and vehicle detection, we propose to employ convolutional neural networks (CNN) which are applied for analyzing visual imagery.[16] Here, we intend to use it for recognizing images of vehicles and non-vehicles and also to locate their position. The problem is modeled as a binary classification task (vehicle/non-vehicle). The model is designed in such a way that it undergoes training by a small sample (e.g., $64 \times 64 \times 3$) coupled with a mono-feature convolutional layer (1×1) at the top, the output of which is counted as probability value for classification.[18-22] Once the model is trained, the input frame's width and height dimensions (width and height) are expanded gradually. As a result of this, the output layer's dimensions map from (1 × 1) to an aspect ratio comparable to that of a new large input. This can be perceived as trimming a large input image into squares of the models' initial input size (64 × 64) and identifying the substances in each of those squares.[11,17]

9.2 LITERATURE STUDY

We study few of the research in the topic of on-road vehicle detection and lane detection as follows:

Sun et al. (2006) presented a survey of vision-based on-road vehicle detection systems which is an important component of a driver-assistance system. He put light on several prominent designed prototypes in the last 15 years. They discussed Hypothesis Generation (HG) methods they are (1) knowledge-based, (2) stereo-based, and (3) motion-based. Edge-based methods, Hypothesis Verification (HV) methods, they are (1) template-based and (2) appearance-based along with critique of each methods. Moreover, effectiveness of optical sensors in detecting on-road vehicle is being discussed. Furthermore, vision-based vehicle detection methods with special references to the monocular and stereovision domains in the last decade have been discussed.[12] In the later time, a concise review has been carried out on vehicle detection by classifying vehicle type classification by processing videos from traffic surveillance cameras.[13]

Song et al. (2019) propose a vision-based vehicle detection system which can be employed for counting vehicles in highway. This research proposes a segmentation approach to uncouple road surface from the image and classifying it into a remote area and a proximal area and subsequently identifying the dimension and location of the vehicle. Next, the Oriented FAST and Rotated BRIEF (ORB) algorithm is employed to locate the vehicle trajectories.[14]

An exhaustive study of the vehicle detection in dynamic conditions such that visual data are processed using a feature representation method known as object proposal methods has been presented by Sakhare et al. (2020).[15] Inspired by the capability and usage of CNN in analyzing a huge image data,[16] Leung et al. (2019) experimented vehicle detection in insufficient and nighttime environment where the objects on photographs are blurry and darkened using deep learning techniques.

9.3 PROPOSED APPROACH

9.3.1 DATASET

The data for investigation are gathered from Udacity which provides a labeled data of 9000 images consisting of vehicles and other 9000 images where vehicles are not present considering all the images are of size (64 × 64). The dataset is an instance of GTI Vehicle Image Database, KITTI Vision Benchmark Suite,[10] and samples are extracted from the project video graphs. A sample of images from the dataset is shown in Figure 9.1.

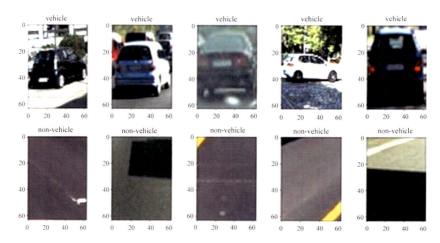

FIGURE 9.1 An abstract view of data of vehicles and non-vehicles.

The data are of 17,760 samples of colored image and image of resolution of (64 × 64) pixels. The dataset has been partitioned into a training set consisting of 90% volume (15,984 samples) and validation set of 10% data (1776 samples) in order to realize a balanced division, which in turn would be a dominant factor later while training and testing the deep learning model and may causes bias toward a particular class.

9.3.2 FLOWCHART

Figure 9.2 shows the detailed procedure involved in vehicle detection while experimenting using CNN.

Vision-Based Lane and Vehicle Detection

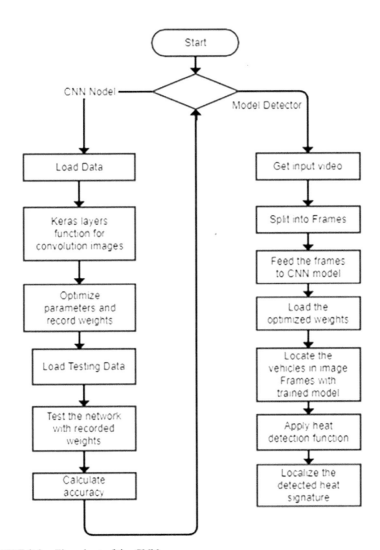

FIGURE 9.2 Flowchart of the CNN process.

9.3.3 ARCHITECTURE

The system and its underlying components are represented in Figure 9.3. The CNN model makes use of Rectified Linear Unit (RELU) activation functions in the convolution layers whereas in order to compute output at output layer, sigmoid function is being utilized. The use of RELU function in hidden layer is attributed to learning which happens in hidden layers. RELU activation function is preferred considering vanishing gradient problem as it is linear for x>0 and 0 for all negative values. Tanh and sigmoid need not be used as activation function for hidden layers because of vanishing gradient problem.

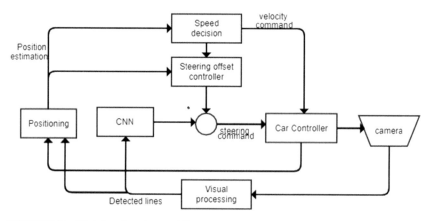

FIGURE 9.3 High level system architecture.

The cross entropy loss function is applied since binary classification and sigmoid activation function are employed in the output layer.

9.4 THE CONVOLUTION OPERATION

Mathematical analysis expresses convolution as a function emanating from the integration of two given functions, such that the shape of one is transformed by superimposition of the other. This is represented in Figure 9.4.

In principle, convolution operation comprises of three elements, they are:

- Input image: 64 × 64 matrices
- Kernel/filter/feature detector: 3 × 3 matrices
- Feature map

Feature map is the byproduct of the integration of input image with feature detector matrix. It is displayed in Figure 9.5.

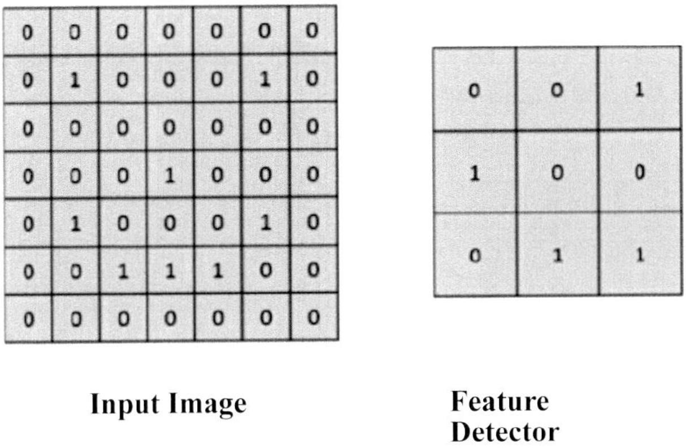

Vision-Based Lane and Vehicle Detection

FIGURE 9.4 Matrix representation.

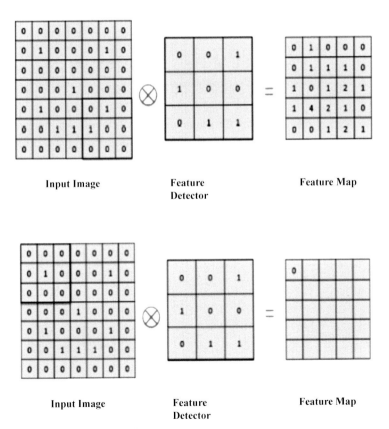

FIGURE 9.5 Construction of feature map.

The generated feature map is employed in the next step in order to trim the input image. It has been exhibited in Figure 9.6 as multiple feature maps form the convolutional layer.

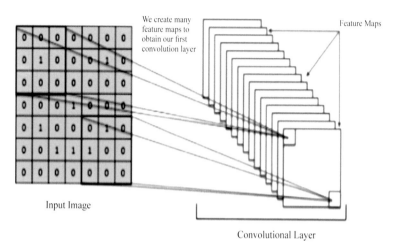

FIGURE 9.6 Representation of a convolutional layer.

The feed of RELU is an additional step in the convolution operation. In order to address nonlinearity in input images, the rectifier linear function is employed (Fig. 9.7).

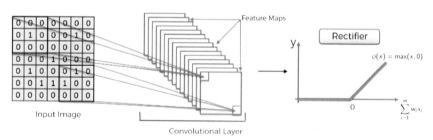

FIGURE 9.7 RELU activation function.

9.5 RESULTS AND DISCUSSION

Detection of lane and vehicles from a video generated from a camera sensor located on a car in motion on a highway is feed as input. Hence, this video can be used in real time on a car to make the car intelligent agent.

9.5.1 VEHICLE DETECTION

The dataset is split into the training set (90%, 15,984 samples) and validation set (10%, 1776 samples).

A neural network is designed to be operated implementing a CNN with an objective to classify the images into car and non-car classes. The fully convolutional network parameters are represented in Table 9.1 which shows the structure of the CNN and its learning parameters. Here, "Conv" represents a convolution layer; all pooling operations are performed using Max_ pooling. The different levels of features of images in both convolution and pooling layer are extracted and it is revealed that 1,347,585 total number of parameters are elicited and trained in training phase.

TABLE 9.1 CNN Parameter Details.

Layer type	Output size	Parameter count
Lambda_1(Lambda)	(64, 64, 3)	0
Conv1 (Conv2D)	(64, 64, 128)	3584
Dropout_1 (Dropout)	(64, 64, 128)	0
Conv2 (Conv2D)	(64, 64, 128)	147,584
Dropout_2 (Dropout)	(64, 64, 128)	0
Conv3 (Conv 2D)	(64, 64, 128)	147,584
Max_pooling2D_1 (Maxpooling2)	(8, 8,128)	0
Dropout_3 (Dropout)	(8, 8,128)	0
Dense1 (Conv2D)	(1, 1,128)	1,048,704
Dropout_4(Dropout)	(1, 1,128)	0
Dense2 (Conv2D)	(1, 1, 1)	129

After training for 20 epochs, the model can be employed for making a prediction on a random sample (Fig. 9.8).

Additionally, the same network trained with our 64 × 64 images can be used to detect cars anywhere in the frame. They scale to whatever the input is, so now we have a heat map output. Consequently, abounding boxes can be drawn on the hot positions.

9.5.2 LANE DETECTION

In this section, we experiment detection of two lane lines on the road for each frame using computer vision techniques (Fig. 9.9).

FIGURE 9.8 Car positioning identification.

FIGURE 9.9 Two lane detection.

9.5.3 PERFORMANCE MEASURE

Test Accuracy and Loss: The following accuracy was obtained on performing classification on the testing data from our dataset (Table 9.2).

TABLE 9.2 Performance Parameters.

Epoch#	Time (sec)	Loss	Accuracy	Val_Loss	Val_Accuracy
1	54	0.0764	0.8940	0.0213	0.9778
2	48	0.0194	0.9756	0.0142	0.9866
3	48	0.0117	0.9855	0.0099	0.9897
4	48	0.0075	0.9904	0.0107	0.9879
5	48	0.0063	0.9923	0.0073	0.9926

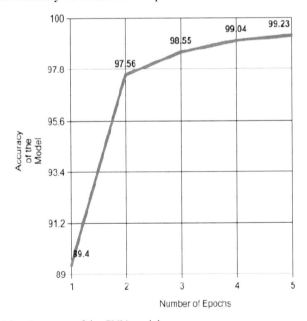

FIGURE 9.10 Accuracy of the CNN model.

It is witnessed from Figure 9.10 that the accuracy of the model increases drastically after the 1st epoch; however, after the 2nd epoch the accuracy increases gradually.

Plot of Loss Function vs Number of Epochs

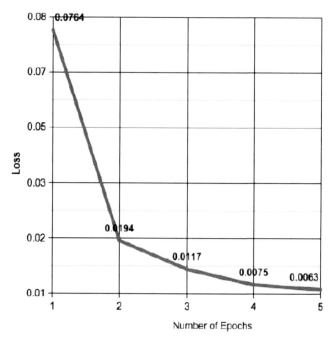

FIGURE 9.11 Loss graph of the model.

In Figure 9.11 above, the value of the loss decreases drastically after the 1st epoch and then decreases gradually after the other 4th epoch.

9.6 CONCLUSION

Most of the literature discussed uses traditional image processing technique for detecting vehicles on the road and also detecting driving lanes and road space from the captured image. However, we employ CNN, a novel technique for detecting lanes and vehicle. The superiority of CNN for the image classification task has been realized, and it delivers the result that was attainable never before. The diversity of current model can be enhanced by training the NN with contrasting and nonidentical images.

KEYWORDS

- **Vehicle Detection**
- **Lane Detection**
- **Autonomous Unmanned Vehicle**
- **Convolutional Neural Networks (CNN)**
- **Automatic driving**

REFERENCES

1. Hadi, R.; Sulong, G.; George, L. Vehicle Detection and Tracking Techniques: A Concise Review. *Signal Image Process. Int. J* **2014**, *5*(1), 1–12.
2. Bertozzi, M.; Broggi, A.; Fascioli, A. Vision-Based Intelligent Vehicles: State of the Art and Perspectives. *Robot. Auton. Syst.* **2000**, *32*, 1–16.
3. Gill, N. K.; Sharma, A. Vehicle Detection from Satellite Images in Digital Image Processing. *Int. J Comput. Intell. Res.* **2017**, *13*(5), 697–705.
4. Chandrasekhar, U.; Das, T. K. A Survey of Techniques for Background Subtraction and Traffic Analysis on Surveillance Video. *Univers. J Appl. Comput. Sci. Technol.* **2011**, *1*(3), 107–113.
5. Liu, M.; Hua, W.; Wei, Q. Vehicle Detection Using Three-Axis AMR Sensors Deployed Along Travel Lane Markings. *IET Intell. Transp. Syst.* **2017**, *11*(9), 581–587.
6. Alletto, S.; Serra, G.; Cucchiara, R. Video Registration in Egocentric Vision Under Day and Night Illumination Changes. *Comput. Vis. Image Underst.* **2017**, *157*, 274–283.
7. Sivaraman, S.; Trivedi, M. Active Learning for On-Road Vehicle Detection: A Comparative Study. *Mach. Vis. Appl.* **2011**, *25*(3), 599–611.
8. Lu, W.; Wang, H.; Wang, Q. A Synchronous Detection of the Road Boundary and Lane Marking for Intelligent Vehicles, Eighth ACIS International Conference on Software Engineering, Artificial Intelligence, Networking, and Parallel/Distributed Computing 2007 IEEE, 2007; pp 741–745.
9. Khalifa, O. O.; Assidiq Abdulhakam, A. M.; Hashim Aisha-Hassan, A. Vision-Based Lane Detection for Autonomous Artificial Intelligent Vehicles, 2009 IEEE International Conference on Semantic Computing, 2009; pp 636–641.
10. Geiger, A. Are We Ready for Autonomous Driving? The Kitti Vision Benchmark Suite, In 2012 IEEE Conference on Computer Vision and Pattern Recognition, 2012; pp 3354–3361. https://doi.org/10.1109/cvpr.2012.6248074.
11. Sun, Z.; Bebis, G.; Miller, R. On-Road Vehicle Detection: A Review. *IEEE Trans. Pattern Anal. Mach. Intell.* **2006**, *28*(5), 694–711.
12. Sivaraman, S.; Trivedi, M. M. A Review of Recent Developments in Vision-Based Vehicle Detection. *IEEE Intelligent Vehicles Symposium (IV)*; Gold Coast, Australia, 2013; pp 310–315.

13. Kul, S.; Eken, S.; Sayar, A. A Concise Review on Vehicle Detection and Classification; *ICET 2017*; Antalya, Turkey, 2017.
14. Song, H.; Liang, H.; Li, H.; Dai, Z; Yun, X. Vision-Based Vehicle Detection and Counting System Using Deep Learning in Highway Scenes. *Eur. Trans. Res. Rev.* **2019,** *11*(51), 1–16.
15. Sakhare, K. V.; Tewari, T.; Vyas, V. Review of Vehicle Detection Systems in Advanced Driver Assistant Systems. *Arch. Comput. Methods Eng.* **2020,** *27*, 591–610.
16. Das, T. K.; Chowdhary, C. L.; Gao, X. Z. Chest X-Ray Investigation: A Convolutional Neural Network Approach. *J Biomim. Biomater. Biomed. Eng.* **2020,** *45*, 57–70.
17. Leung, H. K.; Chen, X. Z.; Yu, C. W.; Liang, H. Y.; Wu, J. Y.; Chen, Y. L. A Deep-Learning-Based Vehicle Detection Approach for Insufficient and Nighttime Illumination Conditions. *Appl. Sci.* **2019,** *9*, 4769.
18. Chowdhary, C. L.; Acharjya, D. P. Segmentation and Feature Extraction in Medical Imaging: A Systematic Review. *Proc. Comput. Sci.* **2020,** *167*, 26–36.
19. Khare, N.; Devan, P.; Chowdhary, C. L.; Bhattacharya, S.; Singh, G.; Singh, S; Yoon, B.. SMO-DNN: Spider Monkey Optimization and Deep Neural Network Hybrid Classifier Model for Intrusion Detection. *Electronics*, **2020,** *9*(4), 692.
20. Reddy, T., Swarna Priya, R. M.; Parimala, M.; Chowdhary, C. L.; Hakak, S.; Khan, W. Z. A Deep Neural Networks Based Model for Uninterrupted Marine Environment Monitoring. *Comput. Commun.* **2020,** *157*, 64–75.
21. Tripathy, A. K.; Das, T. K.; Chowdhary, C. L. Monitoring Quality of Tap Water in Cities Using IoT. In *Emerging Technologies for Agriculture and Environment*, Springer: Singapore, 2020, pp. 107–113.
22. Samantaray, S.; Deotale, R.; Chowdhary, C. L. Lane Detection Using Sliding Window for Intelligent Ground Vehicle Challenge. In *Innovative Data Communication Technologies and Application*, Springer: Singapore, 2021, pp. 871–881.

CHAPTER 10

Damaged Vehicle Parts Recognition Using Capsule Neural Network

KUNDJANASITH THONGLEK[1*], NORAWIT URAILERTPRASERT[2], PATCHARA PATTIYATHANEE[3], and CHANTANA CHANTRAPORNCHAI[3]

[1]Nara Institute of Science and Technology, Nara, Japan

[2]Vidyasirimedhi Institute of Science and Technology, Rayong, Thailand

[3]Kasetsart University, Bangkok, Thailand

*Corresponding author. E-mail: thonglek.kundjanasith.ti7@is.naist.jp

ABSTRACT

An automatic vehicle damage detection platform can enhance the customer claiming process and reduce the unnecessary cost of repair for an insurance company. Typically, the claim estimation process is manual which requires human experts to evaluate the damage cost. This is error-prone, time-consuming, and requires man-hour workers. In this chapter, a damaged vehicle part detection platform, called Intelligent Vehicle Accident Analysis (IVAA) which provides artificial intelligence as a service (AIaaS), is proposed. The system helps automatically assess vehicle parts' damage and severity level. An insurance company can utilize our service to speed up the claiming process. IVAA is built on the docker image which allows the system to be scaled depending on the workload efficiently. Capsule neural network (CapsNet) is applied for damage recognition including two phrases: damage localization and damage classification. The accuracy of the damage localization is 93.28% and the accuracy of the damage classification is 98.47%, respectively.

10.1 INTRODUCTION

The major role of the auto insurance companies is to provide services to their customers supporting the claiming process. Providing the fast services in the field and fast damage repair evaluation is the key success to satisfy their customers. The conventional claiming process usually takes an hour to a day for a customer when the accident happens. For example, he/she has to wait for the arrival of the field personnel, and repair quotation from the insurance experts at the company. The field personnel must spend time to inspect the vehicle at an accident site in the traditional claim process.

Figure 10.1 shows the conventional claiming flow. It starts with an appraisal where either the insurance company will send someone out to the customer car to evaluate the damage, or the customer brings the car to the company or the registered body shop, the car damage is inspected, the fixing process finished, and the reimbursement is done. The whole processing time can be reduced and the customer satisfaction can be increased with the help of artificial intelligence (AI) technology platform.

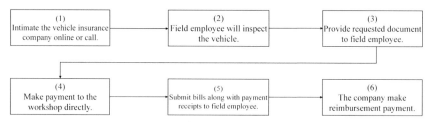

FIGURE 10.1 Traditional clamming process.

There are several core areas in AI such as knowledge, reasoning, problem solving, perception, learning, and ability to manipulate with objects. Deep learning technique is an effective methodology to build an intelligent agent.[1] The area is quite mature in recognition tasks.[2] We apply it to detect damaged parts and damage levels on the vehicle from the accident. The integration of such intelligence into the company service can decrease the claiming process turnaround time and increase the work effectiveness.[21-26]

This chapter focuses on the use of AI in the auto insurance company. The software architecture along with services where the company can utilize on top of its claim process is designed. The prototype application demonstrates how the claim processes can be automated, serving all stakeholders: field worker, car owner, body shop partner to speed up the service anytime and anywhere.

The outline of the chapter is as follows. Next section presents the backgrounds including the literature reviews of car damage evaluation systems and object detection methods. Then, the overall system, the description of each element, and software architecture are presented. The implementation of each system element is then described. Finally, the evaluation process as well as the conclusion remarks are presented.

10.2 BACKGROUND

The research on AI has greatly improved the effectiveness of both manufacturing and service industries. Recent commercial applications to recognize the vehicle accident damage with AI utilized the IBM Watson.[3] Figure 10.2(a) shows an example user interface from IBM Watson. It presents the possible car damage and the types.

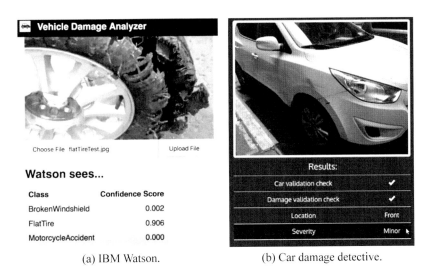

(a) IBM Watson. (b) Car damage detective.

FIGURE 10.2 IBM image recognition software.

In this work, the proposed platform called Intelligent Vehicle Accident Analysis (IVAA) System utilizes images as input data in the same manner as IBM Watson Visual Recognition, and deep learning techniques for recognition. Our work is built upon the integration with open source software, supports multiple image processing at the time and provides a user-friendly and price estimation.

Figure 10.2(b) presents the car damage detective software which is an open source on Github by Neokt.[4] Compared to ours, IVAA is used to detect the specific vehicle part via images, support multiple images of the vehicle, and provide a price estimation on a mobile application.

IBM Watson is a system based on cognitive computing as shown in Figure 10.3. It contains three elements: Watson Visual recognition, web server, and mobile application.

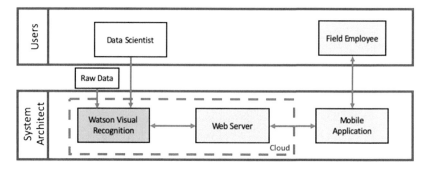

FIGURE 10.3 IBM Watson architecture.

Table 10.1 compares the three softwares in many aspects. The required features of the software are such as classification, localization, automatic model training, and cloud support.

In Ref [3], IBM Watson has on its own recognition engine while ours and car detective are based on Tensorflow and Keras. Compared to these, we can detect more vehicle parts and more damage levels. To achieve an accurate estimation, the model should be able to infer the type of damage. As it affects the expense, it is necessary for the service to suggest the repair or replace the damaged part.

The template matching method is a naive approach for finding a similar pattern in the image.[5] The extension is gray scale-based matching and edge-based matching outlines.[6] The gray scale-based matching is able to reduce the computation time, resulting in up to 400 times faster than the base-line method while edge-based matching performs the matching only on an edge of an object.[7-8] The output is a gray scale image by each pixel representing the degree of matching.

TABLE 10.1 Software's Comparison.

	IVAA	IBM Watson-based [3]	Car damage detective [4]
Features			
Classification	Yes	Yes	Yes
Localization	Yes	No	No
Deep learning library	Tensorflow	IBM Watson	Keras
Store result	Central server	No	No
Deployment	Private cloud	IBM cloud	Private cloud
Labeling system	Yes	No	No
Model training and tuning	API call	IBM Watson	Manual
Interface			
Web application	Yes	Yes	Yes
Mobile platform	via LINE	Native iOS and Android	No
Visualized result			
Parts	Yes	Yes	Yes
Accuracy confident	Yes	Yes	No
Estimate repair price	Yes	No	No
Detection system			
Type of detection	23 parts	3 zones	4 types
Damaged levels	5 levels	3 levels	1 level

Using convolutional neural network (CNN) is another approach for image recognition. It can be used to recognize the category of the image. It can be adopted to perform object detection and localization. Several existing networks are such as the following.

Faster region-based convolutional neural network (R-CNN) is developed based on and provides a user-friendly and price estimation and R-CNNs.[9] The object detection process is separated into two stages.[10] In the first stage, R-CNN applies the selective search to generate the proposed regions. For the second stage, it applies the image classification model to extract features from the proposed region of the previous stage, and then feeds those features to Support Vector Machine (SVM) for generating the final predictions.[11–12]

The improved version of R-CNN is able to provide the faster and more accurate results. The main modifications of Faster R-CNN are to use

CNN to generate the object proposals rather than using selective search in the first stage.[13] This layer is called region proposal network (RPN). RPN uses the base network to extract feature map more precisely from the image. Then, it separates the feature maps to the multiple squared tiles and slides on a small network across each tile continuously. The small network feeds a set of object confidence scores and bounding box coordinates to each location of tile.[14] RPN is designed to be trained in an end-to-end manner. Using Faster R-CNN can reduce the training and detection time.[15–16]

Recently, capsule neural network (CapsNet) has shown a better accuracy than the typical CNN. A capsule is a group of neurons whose activity vector represents the instantiation parameters of a specific type of entity such as an object or an object part.[17] CapsNet contains capsules rather than neurons. The group of capsules learns to detect an object within a given region of the image, and gives the outputs vector which represents the estimated probability that the object is present and whose orientation encodes the object's pose parameters.[18] The capsules are equivariant to the object pose, orientation, and size.

The architecture contains an encoder and a decoder as shown in Figure 10.4. The encoder is used to take the input data and convert it into the n-dimensional vector. The weights of the lower-level capsule (PrimaryCaps) must align with the weights of the higher-level capsule (DigitCaps). At the end of the encoder, an n-dimensional vector is passed to the decoder. The decoder contains many fully connected layers. The main job of the decoder is used to take the n-dimensional vector and attempt to reconstruct from scratch which makes the network more robust by generating predictions based on its own weights.

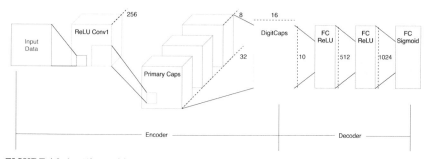

FIGURE 10.4 The architecture of capsule neural network (CapNet) model.[17]

Four main computation stages in a capsule neural are: (1) matrix multiplication, (2) scalar weighting, (3) dynamic routing, and (4) vector-to-vector nonlinearity.

First, the model performs a weight matrix multiplication between the information passed from the higher to the lower layer to encode the information of understanding spatial relationships. Next, the capsules from the lower level adjust its weights according to the weights of the higher level. Dynamic routing algorithm allows the passing data between layers in the network effectively, which increases the time and space complexity. The last step is to compress the information where the condense information can be reused.

10.3 SYSTEM OVERVIEW AND ELEMENTS

There are four user roles in the IVAA system: insurance experts, data scientists, operators, and field employees as shown in Figure 10.5. The four tools are developed for these four users: data labeling tool for insurance experts, deep learning APIs for data scientists, web monitoring application for operators, and LINE chatbot to interact with the back-end server for field employees as in product layers in Figure 10.5.

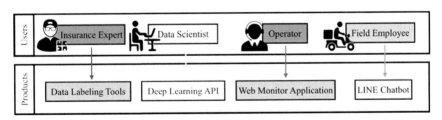

FIGURE 10.5 System architecture.

10.3.1 DATA LABELING TOOLS

The labeling task is one of the time-consuming tasks before the training model process can start. The traditional labeling software such as LabelImg and Imglab[19] works as a standalone application which makes it hard to handle large number of data annotations. Figure 10.6 shows the flows of our tool which has a web interface where the user can collaboratively work on the labeling task. The labeling tool returns a downloadable JSON file for the user for future use. VueJS is used as a frontend framework and REST API server. The labeling tool is also useful for adding more damaged labeled images for future retraining.

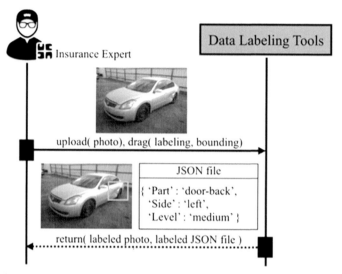

FIGURE 10.6 Data labeling tools sequence diagram.

10.3.2 DEEP LEARNING APIS

APIs are gateways which are designed for data scientists and developers to train and deploy the model. Figure 10.7(a) presents the deep learning API used to input new data and model hyper-parameter for training to create the new deep learning model. The API returns the model identification (model ID) to the user as a link for the model deployment. Figure 10.7(b) shows the testing API which inputs the testing data and model ID to deploy the model. It returns with the list of damaged parts and levels on the vehicle along with the accuracy.

10.3.3 WEB MONITORING APPLICATION

The operators monitor the cases using the web monitoring application. It shows the historical data that contains the number of cases, the number of processed images, and the number of days that system operated. The visualization displays in the heat map style, showing the frequency of accidents by locations and calendar days. Figure 10.8 shows an example of tasks that an office operator monitors with an overview of the case and location.

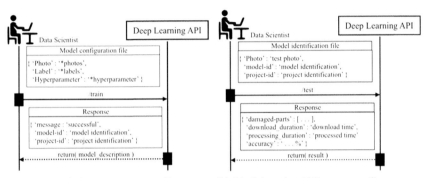

(a) Model training API sequence diagram.   (b) Model testing API sequence diagram.

FIGURE 10.7 Deep learning APIs sequence diagram.

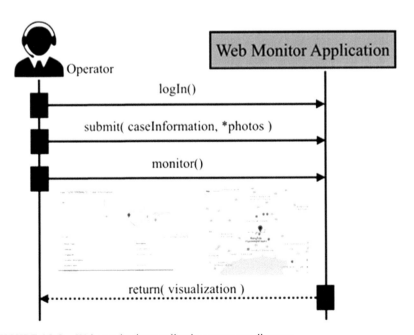

FIGURE 10.8 Web monitoring application sequence diagram.

10.3.4 LINE OFFICIAL INTEGRATION

Field employees use the LINE chatbot service specifically designed for insurance field employees. The chatbot takes the damaged car images via LINE chat and gives the resulting car model and price table images, along with the list of body shop details and locations.

In Figure 10.9, the field employee sends the detail of an accident case, customer ID, shares the accident location, and uploads the damaged car images. Next, the deep learning testing API is executed to recognize the damaged parts and classify the damage level from the submitted photos. The chatbot stores the communication dialogues to the main database.

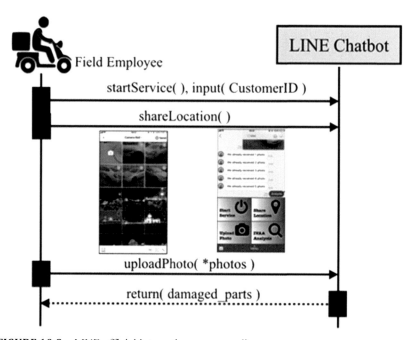

FIGURE 10.9 LINE official integration sequence diagram.

10.3.5 SYSTEM SOFTWARE ARCHITECTURE

All the above services are deployed on the private cloud system with hardware specification listed in Table 10.2. We use the private server to train the model, serving the model, and hosting a website.

Damaged Vehicle Parts Recognition Using Capsule

TABLE 10.2 Hardware Specification.

Hardware	Specification
CPU	Intel (R) Core (TM) i5-2400 CPU @ 3.10 GHz
GPU	NVIDIA Tesla K40c
RAM	24 GB
HDD	4 TB
Internet connection speed	100 Mbps

Figure 10.10 displays the software stack of the system. OpenStack is used for computational resource management such as memory, CPU, network, and other resources to provide for each specified tasks on the container. Visualized resource monitoring part is divided into two sections. The first parts utilize Grafana to monitor resources on the private cloud via OpenStack and the second section utilizes Sahara which monitors resources on the private cloud directly. Kubernetes provides scaling computational resources for each docker container. Docker engine is used visualization container resources to define for task management contribution which is IVAA Core. MongoDB is the main database since it can be scaled out to support more data size and it has a support for unstructured data.

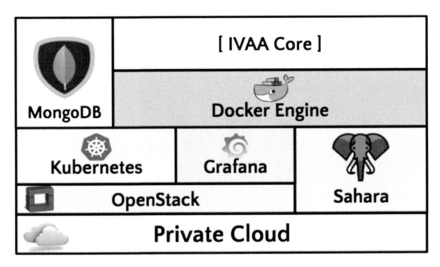

FIGURE 10.10 System software stack.

10.4 IMPLEMENTATION

The implementation is divided into four parts based on four components of IVAA: (1) data labeling tool (2) deep learning APIs (3) web monitoring application, and (4) LINE chatbot.

Figure 10.11 demonstrates the interface of data labeling tools. An insurance expert uploads the damaged vehicle to the IVAA system. The user selects the part and label the photo of a damaged vehicle. The tool makes the labeling process easier than manual labeling. Users can also download the labeled and unlabeled photo data form IVAA system to the local machine.

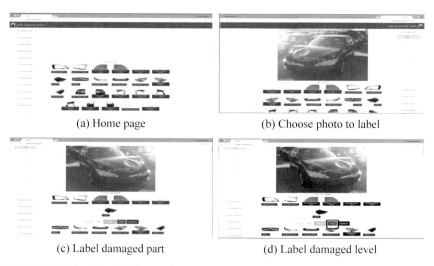

(a) Home page (b) Choose photo to label
(c) Label damaged part (d) Label damaged level

FIGURE 10.11 Data labeling tool.

The labels used for building the models come from multiple insurance experts and the experts may have different subjective opinions on how some of the cases should be labeled. We have studied this scenario by designing a multi-expert learning framework that assumes the information on who labeled the case is available. The framework explicitly models different sources of disagreements and lets us naturally combine labels from different human experts to obtain a consensus classification model representing the model groups of experts converging to and individual expert models.

Labeling data or data annotation is important in deep learning. It provides that initial setup data for the machine learning task. The mislabeled data can lead to wrong prediction easily. Labeling tools are precious and the good ones are usually costly.

In IVAA network, CapsNet is used to recognize the damaged vehicle parts and the levels of severity from the vehicle's photos. We keep each trained model's replica on a single GPU. The memory contains the large number of weights for the layers. In addition, omitting the batch-normalization on top of those layers, we are able to increase the overall number of inception blocks considerably. Table 10.3 shows example damage Levels that are identified from our system.

TABLE 10.3 Representing the Damaged Level.

Damaged level	Colors
No damaged	White
Low level damaged	Yellow
Medium level damaged	Orange
High level damaged	Red
Replacing damaged	Gray

In Figure 10.12, we use IVAA network to recognize many photos obtained from many perspectives. The parts are mapped to images below where the filled color shows the damaged level for each damaged part. Level of damaged parts is based on Thai General Insurance Association (TGIA), which is an organization that promotes and supports the nonlife insurance industry as an accident insurance.

We provide two main APIs for training and testing the model for the data scientists of an insurance company as shown Figure 10.13. The training API requires new images and model configuration as inputs to train and create the new model. It returns the model ID to user for future model usage. Testing API takes testing data and model ID as inputs for testing model. It returns the list of damaged parts and damaged levels on the vehicle image. The APIs conform the REST architectural style or RESTful web services, providing interoperability between computer systems on the Internet. REST-compliant web services allow the requesting systems to access and manipulate textual representations of web resources by using a uniform and predefined set of stateless operations.

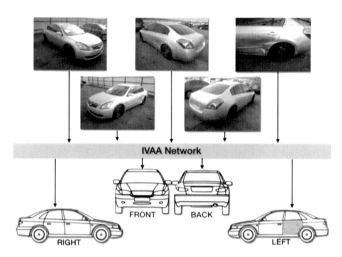

FIGURE 10.12 IVAA network recognizing the photos.

Deep learning APIs are gateways for the user to deploy our system. This enables adding the new data sets and retraining the deep learning model effectively. Incremental retraining allows the increments of model accuracy when having limited computing power.

Figure 10.14 presents the web application developed using VueJS framework with Bulma CSS framework. Web monitoring application is targeted for an office worker, a system administrator and a business manager. An application has six main pages for monitoring and interacting with the system.

The Login page on our web monitoring application is shown in Figure 10.14(a). The Authentication Required function in Go programming language library is adopted. The security in the front end is one way to limit the user interference. However, some users require more flexibility than others and there are always trade-offs.

Figure 10.14(b) shows the dashboard page on our system. It contains three elements: (1) the cases (2) the images processed, and (3) how long systems operated. The first element is the important one where it presents cases reported as well as case management. The second element is about images and their processes. The third element is the system administrative information. Dashboard is a data visualization tool that allows all users to analyze issues to their system. It provides an objective view of performance metrics and serves as an effective foundation for further dialogue.

Damaged Vehicle Parts Recognition Using Capsule

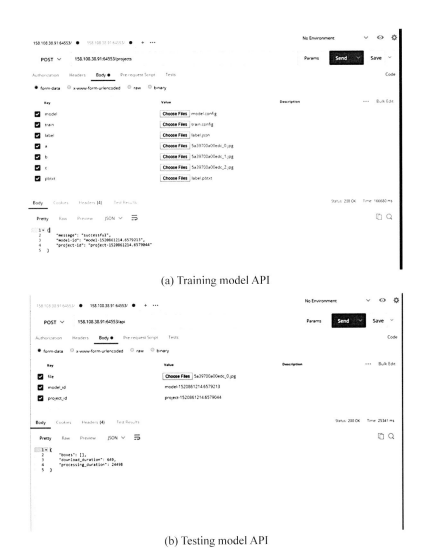

FIGURE 10.13 Deep learning APIs.

Figure 10.14(c) shows the heat map of the cases reported. The primary purpose of heat maps is to visualize the volume events by locations within the data sets and assist in directing viewers toward areas. Fading color shows the density of accident case in that location.

An accident case can be inserted via the case insertion page as shown in Figure 10.14(d) For each accident case, the case identification number, the customer identification number, accident location are required. The images of the damaged vehicle can be uploaded. Drag and drop zone is provided for uploading the photo with convenience. The submitted case is reviewed for approval and the case's information is shown in Figure 10.14(e). In the figure, the case information page is the damaged level on 4 according to the images of the vehicle. Moreover, it indicates the information on the repaired cost based on the damaged level.

The user can find all historical accident cases from the case finder page as shown in Figure 10.14(f). The search can be done by the case identification number, customer identification number, and accident date. The detail button is used to show the detailed information.

Figure 10.15 shows LINE chatbot interface. The LINE official account, namely IVAA as shown in Figure 10.15(a), is supposed for the auto insurance company claiming process. The LINE messaging APIs allows the data to be passed between the server of chatbot application and the LINE platform. When a user sends the chatbot a message, a webhook is triggered

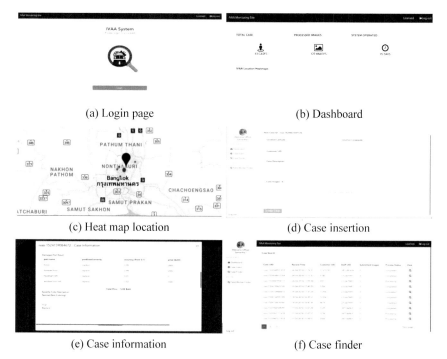

FIGURE 10.14 Web monitoring application.

Damaged Vehicle Parts Recognition Using Capsule 213

(a) Start service (b) Inputting customer ID

(c) Sharing location (d) Uploading photo

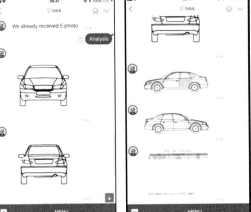

(e) Showing response

FIGURE 10.15 LINE Chatbot.

and the LINE Platform sends a request to the webhook URL. The server sends a request to the LINE platform to respond to the user.

The requests are sent over HTTPS in JSON format. The users can post the IVAA web page onto their Line timelines to make it visible all their friends. The LINE platform allows the user (field employee or customer) to send the damaged car images to the company LINE official account to get the price and damaged results.

After adding IVAA as a friend, the user starts using the service as in Figure 10.15(b). The system requests the customer identification number for authentication. Figure 10.15(c) presents our system authentication to use our service. The service can generate the unique case identification number to the user. The unique case identification number is used for tracking the service progress.

The service also requires the user to share the place of an accident location as shown in Figure 10.15(d) sharing an accident location allows the field worker heading to the location.

Figure 10.15(e) shows our uploading the damaged vehicle's photo in the accident process. At the start, the user takes the photos of the damaged vehicle includes the font side view, the back-side view, the left-side view, and the right-side view. After that, the system acknowledges the receipt of photos. Then, our service returns the analysis result from the deep learning model. The user can visualize the damaged level on the vehicle parts using the difference color as in Figure 10.15(f). In addition, our service can estimate the repair price with the breakdown level damaged parts of vehicle.

10.5 EVALUATION

The evaluation of the application is broken down into three parts. The first part evaluates the IVAA deep learning models. Secondly, the user satisfaction toward web application and LINE chatbot is assessed. Finally, the comparison of the our platform service against the pubic cloud platform is presented.

IVAA deep learning model is compared against the template matching approach and other object detection on the selected car damage data set. Template matching is a technique in digital image processing for finding small parts of an image which matches a template image. The typical object detection algorithm such as R-CNN is used.

IVAA deep learning model utilizes CapsNet to enhance our deep learning model. Due to its recent outstanding performance, we applied CapsNet to detect the damaged vehicle object from the photos, and then

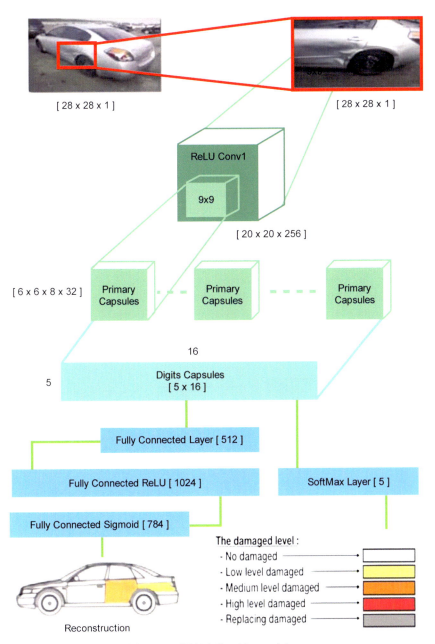

FIGURE 10.16 The architecture of IVAA CapsNet model.

recognize the damaged vehicle parts and the levels of severity. However, since the focus of the work is the application of the model toward the auto insurance claiming process, alternative object detection model is possible.

The architecture of CapNets is shown in Figure 10.16. From the bounding box, the damage part, CapsNet classifies the damage into the mentioned five levels. The part of car is highlighted according to the damage level.

Toyota Camry image set available on https://gitlab.com/Intelligent-Vehicle-Accident-Analysis is used for evaluation. The data set includes 1624 images and we divide 80% training and 20% testing. IVAA utilizing CapsNet yields the accuracy up to 97.21% as shown in Figure 10.17. It has greater accuracy than that of the template matching approach (93.58%). The object detection approach of traditional computer vision technique explores multiple paths where the algorithm is simplified but yet it can achieve higher accuracy with less computation cost (91.53%).

To deploy the model for LINE ChatBot use, we set the threshold for bounding box detection and severe classification to 97.21%. Intersection over under (IoU) for our proposed system is 89.53%. The average inference time per image is 13.12 s on our private cloud. Figure 10.18 implies the inference time when increasing the number of images to 20 images.

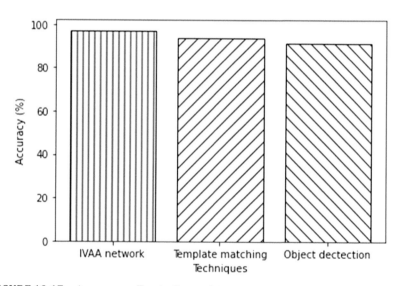

FIGURE 10.17 Accuracy on Toyota Camry data set.

Damaged Vehicle Parts Recognition Using Capsule 217

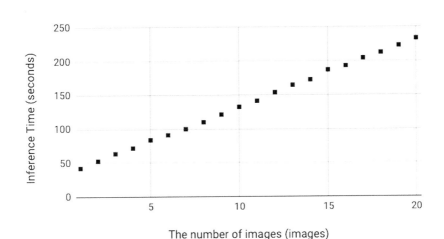

FIGURE 10.18 The inference time per the images.

The confusion matrix is shown in Figure 10.19. Our model can detect the damaged vehicle part very accurately. The data set and the comparison code of the tested car are available at https://gitlab.com/Intelligent-Vehicle-Accident-Analysis.

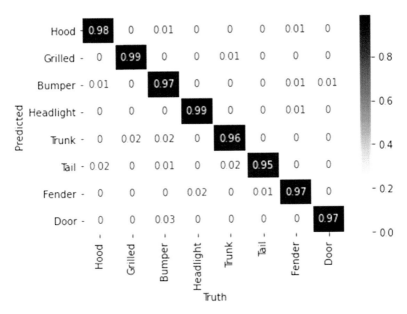

FIGURE 10.19 The confusion matrix of IVAA network.

10.5.1 USER SATISFACTION

The user satisfaction of the application is measured in two aspects: application usage and intelligence module. For the application aspect, the questionnaire asks in the aspect of usability, reliability, security, interface, and availability.[20] The user satisfaction score is shown in Table 10.4.

TABLE 10.4 Usability Test of Application Module (5-Highest).

Aspect	Score
Usability	4.93
Reliability	4.76
Security	4.56
Interface	4.66
Availability	4.56
Average	4.69

For the intelligence module, Table 10.5 shows the summarized score. There are 6 criteria: prediction speed, accuracy, expectation satisfiability, input format satisfiability, and output format satisfiability.

TABLE 10.5 Usability Test of Intelligence Module (5-Highest).

Aspect	Score
Prediction's speed	4.76
Prediction's accuracy	4.56
Prediction's expectation	4.60
Input data format	4.53
Output data format	4.83
Average	4.66

The general opinions from 30 users are collected. The average score for each aspect is shown. The average overall score is 4.69/5 for application side and 4.66/5 for intelligence module. There are 93.3% of users are highly recommend to their friends or companies. Moreover, experience of users expects to use our system in the real situation.

Tables 10.6 and Table 10.7 compare our platform against public services and general web development. IVAA targets at specific task, car damage detection, rather than general vision task. Our service solution using LINE is ready to use and the development process is not complex compared to using WebApp and NativeApp.

TABLE 10.6 Model Platform Comparison.

Feature	IVAA	Google AutoML Vision	Amazon Rekognition
Task	Specific task	General task	General task
Cloud	Private cloud	Public cloud	Public cloud
Custom data	Yes	Yes	No
Custom model	Yes	Yes	No
Car damage detection	Yes	No	No

TABLE 10.7 Development Platform Comparison.

Feature	IVAA	NativeApp	WebApp
Home screen real estate	Low	High	Low
Time to market	Fast	Slow	Middle
Accessibility	LINE	Application	Browser
Security	High	Manual	Manual

10.6 CONCLUSION

IVAA System is one of an artificial intelligence as a service (AIaaS) for an auto-insurance company. The system consists of four modules for four stakeholders: data labeling (for insurance experts), deep learning API for data scientists, the web monitoring application for the operators, and LINE official integration for field employees. We evaluate the system in two aspects: the damage detection capability and the application usability. The accuracy results demonstrate that our object detection model can predict the damage part and damage level correctly up to 97.21% while testing on the Toyota Camry data set. The average image inference time per image is 13.12 seconds. The users are satisfied our system. The average score of user satisfaction is 4.69/5 for application usage and 4.66/5 intelligence module.

Future work includes integrating our system with the driver side application to track the driver location and integrate driving information. The whole process of retraining when adding more images can be automated by periodic schedules. The database of body shops is added to the backend.

KEYWORDS

- **AI as a service**
- **object detection**
- **image classification and localization**
- **capsule neural networks**
- **scalable data processing**

REFERENCES

1. Szegedy, C.; Ioffe, S.; Vanhoucke, V.; Alemi, A. Inception-v4: Inception-Resnet and the Impact of Residual Connections on Learning. International Conference on Learning Representations Workshop, 2016.
2. Ren, S.; He, K.; Girshick, R.; Sun, J. Faster R-CNN: Towards Real-Time Object Detection with Region Proposal Networks. *Adv. Neural Inform. Process. Syst.* **2015**, *28*, 91–99.
3. Madhupriya, G.; Guru, M.; Praveen, S.; Nivetha, B. Brain Tumor Segmentation with Deep Learning Technique. International Conference on Trends in Electronics and Informatics, 2019; pp 758–763.
4. Sutton, R.; Mcallester, D.; Singh, S.; Mansour, Y. Policy Gradient Methods for Reinforcement Learning with Function Approximation. *Adv. Neural Inform. Process. Syst.* **2000**, *12*, 1057–1063.
5. Peters, J.; Schaal, S. Reinforcement Learning of Motor Skills with Policy Gradients. *Int. J Neural Netw.* **2008**, 682–697.
6. Kober, J.; Peters, J. Policy Search for Motor Primitives in Robotics. *Adv. Neural Inform. Process. Syst.* **2009**, *21*, 849–856.
7. Peters, J.; Mulling, K.; Altun, Y. *Relative Entropy Policy Search*; Max-Planck Gesellschaft: Menlo Park, 2010; pp 1607–1612.

8. Kim, Y.; Araujo, S. Grayscale Template Matching Invariant to Rotation, Scale, Translation, Brightness, and Contrast. *Adv. Image Video Technol.* **2007,** 100–113.
9. Hofhauser, A.; Steger, C.; Navab, N. Edge-Based Template Maching and Tracking for Perspectively Distorted Planar Objects. *Adv. Visual Comput.* **2008,** 35–44.
10. Stulp, F.; Sigaud, O. Path Integral Policy Improvement with Covariance Matrix Adaption. Proceedings of International Conference on Machine Learning, 2012.
11. Kohl, N.; Stone, P. Policy Gradient Reinforcement Learning for Fast Quadrupedal Locomotion. International Conference on Robotics and Automation, 2004; pp 2619–2624.
12. Arthayakun, S.; Kamonsantiroj, S.; Pipanmaekaporn, L. Image Based Classification for Warm Cloud Rainmaking Using Convolutional Neural Networks. International Conference on Computer Science and Software Engineering, 2018.
13. Phaudphut, C.; Soin, C.; Phusomsail, W. A Parallel Probabilistic Neural Network ECG Recognition Architecture Over GPU Platform. International Joint Conference on Computer Science and Software Engineering, 2016; pp 1–7.
14. Toncharoen, R.; Piantanakulchai, M. Traffic State Prediction Using Convolutional Neural Network. International Joint Conference on Computer Science and Software Engineering, 2018; pp 1–6.
15. Kongkhaensarn, T.; Piantanakulchai, M. Comparison of Probabilistic Neural Network with Multilayer Perceptron and Support Vector Machine for Detecting Traffic Incident on Expressway Based on Simulation Data. International Joint Conference on Computer Science and Software Engineering, 2018; pp 1–6.
16. Huang, J.; Rathod, V.; Sun, C.; Zhu, M.; Korattikara, A.; Fathi, A.; Fisher, I.; Wojna, Z.; Song, Y.; Guadarrama, S.; Murphy, K. Speed/Accuracy Trade-Off for Modern Convolutional Object Detectors. International Conference on Computer Vision and Pattern Recognition, 2017; pp 3296–3305.
17. Deisenorth, P.; Rasmussen, E. A Model-Based and Data Efficient Approach to Policy Search. International Conference on Machine Learning, 2011; pp 465–472.
18. Sabour, S.; Frosst, N.; Hinton, E. Dynamic Routing Between Capsules. *Adv. Neural Inform. Process. Syst.* **2017,** *30*, 3856–3866.
19. Mallea, G.; Meltzer, P.; Bentley, J. Capsule Neural Networks for Graph Classification Using Explicit Tensorial Graph Representations; 2019.
20. Sommerville, I. *Software Engineering*; Addison Wesley: Harlow, 2010.
21. Chowdhary, C. L.; Goyal, A.; Vasnani, B. K. Experimental Assessment of Beam Search Algorithm for Improvement in Image Caption Generation. J. Appl. Sci. Eng. **2019,** *22*(4), 691–698.
22. Khare, N.; Devan, P.; Chowdhary, C. L.; Bhattacharya, S.; Singh, G.; Singh, S.; Yoon, B. SMO-DNN: Spider Monkey Optimization and Deep Neural Network Hybrid Classifier Model for Intrusion Detection. *Electronics,* **2020,** *9*(4), 692.
23. Chowdhary, C. L.; Acharjya, D. P. Segmentation and Feature Extraction in Medical Imaging: A Systematic Review. *Proc. Comput. Sci.*, **2020,** *167,* 26–36.
24. Reddy, T.; Swarna Priya, R. M., S. P.; Parimala, M.; Chowdhary, C. L.; Hakak, S.; Khan, W. Z. A Deep Neural Networks Based Model for Uninterrupted Marine Environment Monitoring. Comput. Commun. **2020,** *157,* 64–75

25. Reddy, G. T.; Bhattacharya, S.; Ramakrishnan, S. S.; Chowdhary, C. L.; Hakak, S.; Kaluri, R.; Reddy, M. P. K. (2020, February). An Ensemble based Machine Learning model for Diabetic Retinopathy Classification. In 2020 International Conference on Emerging Trends in Information Technology and Engineering (ic-ETITE) (pp. 1-6). IEEE.
26. Shynu, P. G.; Shayan, H. M.; & Chowdhary, C. L. (2020, February). A Fuzzy based Data Perturbation Technique for Privacy Preserved Data Mining. In 2020 International Conference on Emerging Trends in Information Technology and Engineering (ic-ETITE) (pp. 1-4). IEEE.

CHAPTER 11

Partial Image Encryption of Medical Images Based on Various Permutation Techniques

KIRAN[1*], B. D. PARAMESHACHARI[2], H. T. PANDURANGA[3], and ROCÍO PÉREZ DE PRADO[4]

[1]Dept. of ECE Engineering, Vidyavardhaka Engg. College, Mysuru, India

[2]GSSS Institute of Engineering & Technology for Women, Mysuru, India

[3]Dept. of ECE Engineering, Govt. Polytechnic, Turvekere, Tumkur, India

[4]Linares School of Engineering, Telecommunication Engineering Department, Scientific-Technical Campus of Linares-University Ave. Linares (Jaén), Spain

*Corresponding author. E-mail: kiran.mtech12@gmail.com

ABSTRACT

Medical image security becomes more and more important. Full image encryption is not necessary in the field of medical because partial amount of encryption is enough to provide the security. Here Proposed is a partial image encryption of medical images, which uses different permutation techniques. Proposed technique mainly consists of permutation and diffusion process. Original medical image divided into nonoverlapping blocks with the help of block size table. Then position of each pixel in every blocks are shuffled according to chaotic sequence generated from the chaotic map system and predefined block size table. In the diffusion process, based on basic intensity image (BII) and different permutation technique, the mapping operation apply to get partially encrypted medical images. Experiential results show that proposed method provide more security with less complexity and computational time.

11.1 INTRODUCTION

In the last two decades, owing to rapid progress in communication systems and multimedia technology, digital image encryption has played an important role in secure communication applications in the fields of military, medical, satellite, and so on. Partial encryption is one of the most commonly used encryption technique in the medical field where it is not necessary to encrypt the full image. Because small amount of encryption leads to high security and less computational in terms of time and complexity. In partial image encryption, generally it gives some clue about original image.[2,9,10,11]

Guodong et al.[18] explained about auto-blocking technique for segmenting the original image with predefined block size and ECG signal used as key for generating random sequence using chaotic system. Based on different ECG signals, different key is obtained for encrypting the different images. Lu and Gou et al.[17] explained about the segmentation of image done by using different block size for permutation process and for diffusion process implemented with the help of dynamic index technique. Permutation of image is done by either horizontal or vertical cross section of the original image. Zhongyunhua et al.[5] introduced a two-dimensional sine logistic map based image encryption algorithm. It has a lot of advantages as compared to chaotic map like greater ergodicity, hyper chaotic property, and low of the implementation is very low. Panduranga et al.[7] describes the partial image encryption scheme for controlling the amount of encryption with a different step size. In this method, multistage hill cipher technique is used for manipulating the pixels value in the original image and division of blocks are varied to control the amount of encryption. This method can be used in the smart cameras where the specific amount of encryption is required. Kumar et al.[8] proposed block-wise approach for encrypting image partially. Where different combination of block size produce various encrypted partial images. To shuffle the pixels within group, chaotic system has been incorporated.[23-26]

Xiangyun et al.[14] introduced the concept of color image encryption in spatial and frequency domain which includes discrete wavelet transform (DWT) and six-dimensional hyper chaos. In spatial domain, key sequence generated by hyper chaotic and segmentation of original image is done with the help of DWT which leads to four frequency band of original image in frequency domain. Belaze et al.[2] explained about most common shuffling-diffusion process based on an image encryption system where the diffusion of the image occurred first then followed by chaos-based shuffling process. Xiang et al. describe the medical image full and selective image encryption. This technique consists of several stages where every stage consists of permutation phase and diffusion phase. Block-based concept is used to permute and encrypt with the help of chaotic map.[16] Parameshachari et al. (2013) proposed partial encryption for medical images which uses the DNA encoding and addition techniques. Random image is generated from chaotic map which undergo DNA addition with original image to get different partial encrypted images.[10] Bhatnagar and Wu explain the concept of SVD and pixel of interest to encrypt selectively the group of pixels in the input image. The idea of this method is to use saw tooth space fiand Q curve to shuffle the pixel positions and diffusion can be done with the help of nonlinear chaotic map.[3] Mahmood and Dony explained algorithm which divides the medical image into two parts based on amount of significant and nonsignificant information namely the region of interest (ROI) and the region of background (ROB). To reduce the encryption time, AES applied to ROI and Gold code (GC) to ROB.[6] Parameshachari et al. introduced the partial encryption of color RGB image. In this method, input color image is segmented into number of macroblocks. Based on the interest, few significant blocks are selected and encrypted using chaotic map.[9] Chowdhary et al.[19] explained about different fuzzy segmentation methods used for dividing and detecting brain tumors in the medical MRI images. Chowdhary et al.[20] introduced a hybrid scheme for breast cancer detection using intuitionistic fuzzy rough set technique. The hybrid scheme starts with image segmentation using intuitionistic fuzzy set to extract the zone of interest and then to enhance the edges surrounding it. Chowdhary[21] explained about how clustering approach holds the positive points of possibilistic fuzzy c-mean that will overcome the coincident cluster problem, reduce the noise, and bring less sensitivity to an outlier. Chowdhary et al.[22] explained experimental assessment of beam search algorithm for improvement in image caption generation.

The entire chapter is divided into various sections, where Section 2 explains about various permutation methods used in the proposed system. Section 3 gives detailed description of proposed partial encryption system based on various permutation techniques. Performance metric analysis of the proposed system is explained in Section 4. At last conclusion of the chapter is described in Section 6.

11.2 PERMUTATION TECHNIQUES

Generally any image encryption algorithm involves the permutation and diffusion process. In the permutation, where the position of the pixels changes there are different techniques used for permuting the image, which includes chaotic map continuous chaos (CC), Gray code (GC), Sudoku code (SC), and Arnold cat map (AC). Detailed description of every permutation method is explained below.

11.2.1 CHAOTIC MAP

Selecting the map for any image encryption scheme is very important and also the major step. Here, chaotic map has been used for permutation process because of its tremendous features like periodic windows, chaotic interval, complexity, sensitivity to initial condition, uses of chaotic system in encryption system more secure and less complex. Chaotic map fulfills the requirement of encryption system in terms of privacy and efficiency.[10]

Mathematical chaotic map can be defined by using following eq 11.1 which includes two parameters that is r and x0 that will be considered as key for the encryption.

$$X_{n+1} = r * X_n (1 - X_n) \qquad (11.1)$$

where the range of initial parameter x lies between 0 and 1. The range of r lies between 3.57 and 4.

11.2.2 CONTINUOUS CHAOS (CC)

Another system used for generating the random sequence is the continuous chaotic system which can be defined by Lorenz system[30] as shown in eq 11.2.

$$\begin{bmatrix} x' \\ y' \\ z' \end{bmatrix} = \begin{bmatrix} -10 & 10 & 0 \\ 8 & 4 & 0 \\ 0 & 0 & -8/3 \end{bmatrix} \begin{bmatrix} x \\ y \\ z \end{bmatrix} + \begin{bmatrix} 0 \\ -xz \\ xy \end{bmatrix}. \qquad (11.2)$$

To remove the near predictability of above CC system by adjusting the output sequences x, y, z. Later long sequence can be obtained by combining every values of all the three sequences. This sequence is arranged in the nondecreasing order and store the new index values for the shuffling process.

11.2.3 GRAY CODE (GC)

GC technique[31] is simple and a more effective permutation method, which is defined in eq 11.3.

$$G = B \oplus (B \gg (q + 1)) \qquad (11.3)$$

where B indicates the k-bit number, G is the k-bit GC value, $\oplus$ is the binary exclusive OR (XOR) operation, q is an integer, and $\gg$ is the binary right shift. The GC for a k-bit number is a also a k-bit number.

To shuffling the original image using GC, firstly image has to be converted into a row array of pixels. Let us consider an example where GC uses four numbers P1, P2, off1, off 2. It should be mentioned that off1 and off 2 are k-bits numbers. For each pixel location, two GC values X1 and X2 are calculated, where X1 = Gray(A, P 1) $\oplus$ off 1 and X2 = Gray(A, p2) $\oplus$ off 2. Then, read the pixel at location X1 and place it in location X2 in the permuted image.

11.2.4 SUDOKU CODE (SC)

One of the most commonly used permutation method is sudoku where every row contains same number of elements but in a different order. Similarly, column also contains same number of elements but in a different order. The name "Sudoku Code" was inspired by mathematical papers by Leonhard Euler.[29]

Algorithm for SC is described below:

Algorithm: Sudoku Code Generation S=Sudoku (p1,p2)

Require: p1 and p2 are two length-Q sequence
Ensure: Pisasudoku Code of order K
1.Kseed=sorting(p1)
2.Kshift=sorting(p2)
3.for i=0 to N-1 do
4. do<=S(i,:)=circularshift(Kseed, Kshift(i))
5: end for

11.2.5 ARNOLD CAT MAP (AC)

AC is one of the important random shuffling method[28] which is defined by following eq 11.4. It consists of p and q positive integer and can be considered as key.

$$\begin{bmatrix} s' \\ r' \end{bmatrix} = \begin{bmatrix} 1 & p \\ q & 1+pq \end{bmatrix} \begin{bmatrix} s \\ r \end{bmatrix} \quad (11.4)$$

where (s, r) and (s, r) are the picture coordinates of the input and permuted image, respectively. Figure 11.1 shows how various permuted images are obtained by using above permutation techniques.

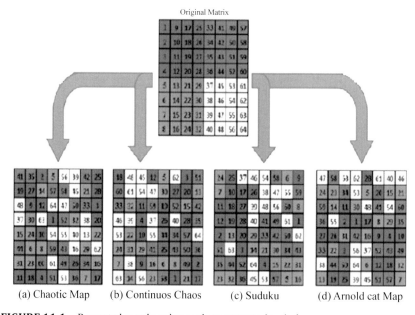

(a) Chaotic Map (b) Continuos Chaos (c) Suduku (d) Arnold cat Map

FIGURE 11.1 Permuted matrix using various permuted techniques.

11.3 PROPOSED PARTIAL IMAGE ENCRYPTION (PIE) METHOD

Architecture of proposed partial encryption scheme using various permutation techniques is shown in Figure 11.2 where it consists of permutation stage followed by mapping stage. At fi3.W medical input image whose size should in-terms of powers of 2 segmented into nonoverlapping macroblock and size of macroblock has been defined in Table 11.1. By using one of the abovementioned permutation technique especially chaotic map used for changing the pixels within every segmented block to get the various intermediate permuted images. With the use of basic intensity image (BII) where it contains all the pixels ranging from 0 to 255 in mapping process along with one of the permutation method to get the various partial encrypted images. The detailed description of block-wise permutation and mapping process can be explained as mentioned below. Permutation process:

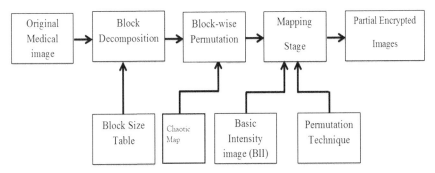

FIGURE 11.2 Architecture of partial image encryption system for medical images.

The steps explain about how input image is permuted by using chaotic map system.

Step 1: Input plain medical image having a size M * N.
Step 2: Partitioning the plain medical image into nonoverlapping macroblocks according to predefined block size from block size Table 11.1.

TABLE 11.1 Block Size List.

Sl. no	Block size
1	4 × 4
2	8 × 8
3	16 × 16
4	32 × 32
.	.
.	.
.	.
N	(n/2) × (n/2)

Step 3: Generate the random sequence with the help of chaotic system eq 11.1 along the initial key x0 and r. Chaotic sequence X can be represented as:

$$X = x1, x2, x3, \ldots\ldots\ldots\ldots\ldots\ldots xn - 1$$

Step 4: Arrange the above chaotic sequence X in the increasing order and store the newly obtained index values.

Step 5: With respect to new position values, randomly permute the position of gray values in every block.

Step 6: To get the randomly permuted image by merging all the blocks in a nonoverlapped fashion to obtain permuted image.

After block-wise permutation, we get different permuted images. Apply the different permutation methodology for the permuted images in the mapping stage process and select different permutation techniques. Steps involving in mapping stage are as follows:

Step 1: Input BII for the mapping process along with one of the permutation technique.

Step 2: Every pixel of intermediate permuted image to be converted into its binary 8-bit number.

Step 3: Split the binary 8-bit binary into two 4-bit number by grouping most significant 4-bit as a higher nibble and least significant 4-bit as a lower nibble.

Step 4: Upper and lower nibble 4-bit number converted into its equivalent decimal value.

Step 5: By using two decimal values obtained from step 4 are used to fetch the gray value pixel basic intensity mapping image. Where decimal value of a upper nibble is treated as a row indicator and decimal value of lower nibble is treated as a column indicator for mapping image.

11.4 PERFORMANCE METRIC FOR PROPOSED PARTIAL IMAGE ENCRYPTION SCHEME

To know the performance of proposed partial encryption system, the following performance metrics has been used for evaluation purpose.

11.4.1 MEAN SQUARE ERROR (MSE)

MSE is calculated between original image and encrypted image obtained from the proposed system which gives average squared difference between original and encrypted image. Mathematically MSE can be defined by following eq 11.5.[15]

$$MSE = \frac{1}{M*N}\sum_{i=1}^{N}\sum_{j=1}^{w}[org(i,j) - enc(i,j)]^2 \quad (11.5)$$

where M, N is the total row and total column of image and $org(i,j)$ is original input image and $enc(i,j)$ is encrypted image.

11.4.2 PEAK SIGNAL-TO-NOISE RATIO (PSNR)

PSNR is inversely proportional to MSE. PSNR refects the encryption quality. Mathematically PSNR can be defined as eq 11.6.[1]

$$PNSR = 20*log_{10}\left[\frac{255}{MSE}\right] \quad (11.6)$$

where MSE is mean square error between input original image and encrypted image and can be obtained by using eq 11.5.

11.4.3 NPCR AND UACI

Number of pixels change rate (NPCR) is generally calculated between original input image and encrypted image. Where NPCR indicates how many pixels in the original image change with respect to encrypted image. Higher the NPCR greater the security and more the encryption. Mathematically, NPCR is defined by following eq 11.8. Unified average changing intensity (UACI) which is inversely related to NPCR. UACI gives average changing intensity values in the original image.[15] Mathematically, UACI can be calculated by using eq 11.7.

$$= \frac{1}{M*N} \sum_{i,j} \frac{\text{UACI} |org(i,j) - enc(i,j)|}{255} \quad (11.7)$$

$$\times 100\%$$

where, M stands for image's width, N stands for image's height, and where $D(i,j)$ is defined as follows (Table 11.2):

$$D(i,j) = \begin{cases} 1 & \text{if } I(i,j) \neq E(i,j); \\ 0 & \text{if } I(i,j) = E(i,j), \end{cases}$$

where $I(i,j)$ and $E(i,j)$ are the original input image and output cipher image, respectively.

11.4.4 UNIVERSAL IMAGE QUALITY (UIQ) INDEX

Universal index quality is used for calculating similarity between original image and cipher image. Range of UIQ is [−1,1] where value 1 indicates more similarity and value −1 indicates less similarity. UIQ is defined as follows:[13]

$$UQI(x,y) = \frac{\sigma_{xy}}{\sigma_x \sigma_y} * \frac{2\mu_x \mu_y}{\mu^2_x + \mu^2_y} * \frac{2\sigma_x \sigma_y}{\sigma^2_x + \sigma^2_y} \quad (11.9)$$

where μx, μy, σx, σy, and σxy are the mean of x and y, variance x and y, and the covariance of x and y, respectively.

11.4.5 STRUCTURAL SIMILARITY INDEX MEASURE (SSIM)

The SSIM is the extended version of the UIQ index. Range of SSIM is [−1,1] where value 1 indicates more similarity and value −1 indicates less similarity. SSIM is defined as follows:[12]

$$SSIM(x,y) = \frac{(2\mu_x \mu_y + C1)(2\sigma_{xy} + C2)}{(\mu^2_x + \mu^2_y + C1)(\sigma^2_x + \sigma^2_y + C2)} \quad (11.10)$$

$$MSSIM = \frac{1}{M}\sum_{j=1}^{M} SSIM(x_j, y_j) \quad (11.11)$$

where *C1*, *C2* are two constants and are used to stabilize the division with weak denominator.

TABLE 11.2 Results Obtained from Proposed Method for Baby Image.

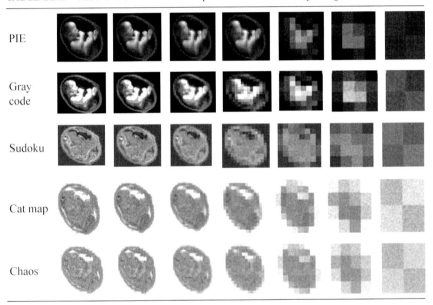

11.5 EXPERIMENTAL RESULTS

In this experiment, we take different images of size 512 × 512. Results of proposed method are tabulated in Tables (from 11.3 to 11.8). From Tables 11.3 and 11.4, we come to know that amount of encryption in terms of MSE and PSNR is less for the combination of GC mapping and permutation with lower block size (4 × 4) and more for the combination of CC mapping and permutation with lower block size. There is no specific control over the amount of encryption in terms of MSE/PSNR but we can vary the amount of MSE/PSNR by choosing appropriate block size and permutation techniques. From Table 11.5, we come to know that NPCR is less for the combination of CC mapping and permutation with lower block size (4 × 4) and almost more than or equal to 99 for GC mapping. From Table 11.6, UACI is less for the combination of GC mapping and permutation with lower block size and varies for remaining combinations. From Tables 11.7 and 11.8 SSIM and UQI more for the combination of GC mapping and permutation with lower block size and varies for the other combinations (Tables 11.7–11.11).

TABLE 11.3 MSE for Baby Image for Different Permutation Techniques.

	MSE for Baby Image						
PIE List	1	2	3	4	5	6	7
GC	31.62	35.21	40.24	47.39	56.72	68.85	82.45
SC	33.10	32.72	32.38	32.40	32.72	33.18	39.19
AC	28.43	27.84	26.97	26.16	25.00	22.35	15.67
CC	26.37	26.53	26.68	26.50	25.38	23.45	17.17

TABLE 11.4 PSNR for Baby Image for Different Permutation Techniques.

	PSNR for Baby Image						
PIE List	1	2	3	4	5	6	7
GC	33.13	32.66	32.08	31.37	30.59	29.75	28.96
SC	32.93	32.98	33.02	33.02	32.98	32.92	32.19
AC	33.59	32.68	33.82	33.95	34.15	34.63	36.17
CC	33.91	33.89	33.86	33.89	34.08	34.42	35.78

TABLE 11.5 NPCR for Baby Image for Different Permutation Techniques.

	NPCR for Baby Image						
PIE List	1	2	3	4	5	6	7
GC	54.92	58.57	60.46	62.29	66.76	73.31	81.12
SC	99.65	99.63	99.64	99.60	99.54	99.53	99.47
AC	99.80	99.78	99.79	99.81	99.83	99.84	99.83
CC	99.71	99.75	99.77	99.78	99.80	99.79	99.79

TABLE 11.6 UACI for Baby Image for Different Permutation Techniques.

	UACI for Baby Image						
PIE List	1	2	3	4	5	6	7
GC	4.90	5.73	6.94	8.54	11.13	14.45	19.58
SC	29.54	29.57	29.63	29.72	29.90	30.31	32.01
AC	63.40	63.42	63.39	63.36	63.22	62.77	61.30
CC	57.98	57.98	58.06	58.09	57.99	57.75	56.23

TABLE 11.7 SSIM for Baby Image for Different Permutation Techniques.

	SSIM for Baby Image						
PIE List	1	2	3	4	5	6	7
GC	0.7426	0.6718	0.5356	0.4110	0.2572	0.1394	0.0271
SC	0.0166	0.0184	0.0180	0.0177	0.0156	0.0159	0.0152
AC	0.0122	0.0093	0.0124	0.0117	0.0129	0.0100	0.0095
CC	0.0185	0.0174	0.0138	0.0116	0.0141	0.0115	0.0113

TABLE 11.8 UQI for Baby Image for Different Permutation Techniques.

	UQI for Baby Image						
PIE List	1	2	3	4	5	6	7
GC	0.9287	0.8612	0.7488	0.6186	0.5769	0.4109	0.3439
SC	0.2376	0.2372	0.2391	0.2458	0.2606	0.2760	0.2795
AC	0.2435	0.2427	0.2399	0.2341	0.2215	0.2093	0.1881
CC	0.2479	0.2475	0.2449	0.2397	0.2307	0.2155	0.2005

TABLE 11.9 Gray Code (GC) Results Obtained from Proposed Method for Lena and Pepper Images.

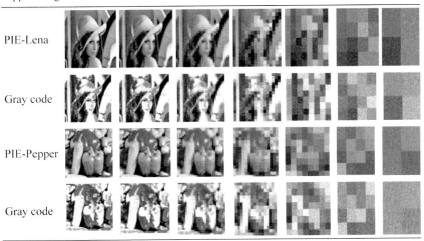

TABLE 11.10 NPCR and UACI Comparison Between Proposed GC Code Method and Existing Method.

Images	GC code		Ref. [11]	
	NPCR	UACI	NPCR	UACI
Lena	99.59	29.01	98.69	18.23
Pepper	99.62	29.97	97.23	22.21

TABLE 11.11 MSE and PSNR Comparison Between Proposed GC Code Method and Existing Method.

Images	GC code		Ref. [11]	
	MSE	PSNR	MSE	PSNR
Lena	89.29	28.62	9.83	6801
Pepper	94.26	28.38	9.10	8051

11.6 CONCLUSION

Here, proposed is a partial image encryption of medical images, which uses various permutation techniques. Proposed system mainly consists of permutation and diffusion process. Where the block-wise permutation of image is done by using chaotic system and with the help of block size table. In diffusion stage, mapping process is used for altering the pixel values. From the experiment, result shows that amount of encryption varies from different permutation techniques. Based on the requirement in the application, a particular permutation-based partial encryption technique can be used. The advantages of proposed method is less complexity and less computation time.

ACKNOWLEDGMENT

This work is supported by VTU Belagaum, Karnataka and Electronics and communication Research centre, GSSSIETW, Mysuru, Karnataka.

KEYWORDS

- **permutation**
- **encryption**
- **chaotic map**
- **intensity image**
- **arnold map**
- **Gray code**

REFERENCES

1. Ahmad, J.; Ahmed, F. Efficiency Analysis and Security Evaluation of Image Encryption Schemes. *Computing* **2010**, *23*, 25.
2. Belazi, A.; Abd El-Latif, A. A.; Belghith, S. A Novel Image Encryption Scheme Based on Substitution-Permutation Network and Chaos. *Signal Process*. **2016**, *128*, 155–170.
3. Bhatnagar, G.; Jonathan Wu, Q. M. Selective Image Encryption Based on Pixels of Interest and Singular Value Decomposition. *Digital Signal Process.* 22(4), 648–663, 2012.
4. Goel, A.; Chaudhari, K. Median Based Pixel Selection for Partial Image Encryption, 2015.
5. Hua, Z.; Zhou, Y.; Pun, C. M.; Philip Chen, C. L. 2d Sine Logistic Modulation Map for Image Encryption. *Inform. Sci*. *297*, 80–94, 2015.
6. Mahmood, A. B.; Dony, R. D. Segmentation Based Encryption Method for Medical Images. In 2011 International Conference for Internet Technology and Secured Transactions, 2011; pp 596–601.
7. Naveenkumar, S. K.; Panduranga, H. T.; et al. Partial Image Encryption for Smart Camera. In Recent Trends in Information Technology (ICRTIT), 2013 International Conference on, 2013; pp 126–132.
8. Panduranga, H. T.; Naveenkumar, S. K.; et al. Partial Image Encryption Using Block Wise Shuffling and Chaotic Map. In Optical Imaging Sensor and Security (ICOSS), 2013 International Conference on, 2013; pp 1–5.
9. Parameshachari, B. D.; Karappa, R.; Sunjiv Soyjaudah, K. M.; Devi KA, S. Partial Image Encryption Algorithm Using Pixel Position Manipulation Technique: The Smart Copyback System. In 2014 4th International Conference on Arti_cial Intelligence with Applications in Engineering and Technology, 2014; pp 177–181.
10. Parameshachari, B. D.; Panduranga, H. T.; Naveenkumar, S. K.; et al. Partial Encryption of Medical Images by Dual Dna Addition Using Dna Encoding. In 2017 International Conference on Recent Innovations in Signal processing and Embedded Systems (RISE), IEEE, 2017; pp 310–314.
11. Som, S.; Mitra, A.; Kota, A. A Chaos Based Partial Image Encryption Scheme, 2014.
12. Wang, Z.; Bovik, A. C. Modern Image Quality Assessment. *Synth. Lect. Image Video Multimedia Process.* **2006**, *2*(1), 1–156.
13. Wang, Z.; Bovik, A. C.; Sheikh, H. R.; Simoncelli, E. P. Image Quality Assessment: from Error Visibility to Structural Similarity. *IEEE Transac. Image Process.* **2004**, *13*(4), 600–612.
14. Wu, X.; Wang, D.; Kurths, J.; Kan, H. A Novel Lossless Color Image Encryption Scheme Using 2d dwt and 6d Hyperchaotic System. *Inform. Sci.* **2016**, *349*, 137–153.
15. Wu, Y.; Noonan, J. P.; Agaian, S. Npcr and UACI Randomness Tests for Image Encryption. Cyber Journals: Multidisciplinary Journals in Science And Technology, Journal of Selected Areas in Telecommunications (JSAT), 2011; pp 31–38.
16. Xiang, T.; Hu, J.; Sun, J. Outsourcing Chaotic Selective Image Encryption to the Cloud with Steganography. *Digital Signal Process*. **2015**, *43*, 28–37.

17. Xu, L.; Gou, X.; Li, Z.; Li, J. A Novel Chaotic Image Encryption Algorithm Using Block Scrambling and Dynamic Index Based di_usion. *Opt. Lasers in Eng.* **2017,** *91*, 41–52.
18. Ye, G.; Huang, X. An Image Encryption Algorithm Based on Autoblocking and Electrocardiography. *IEEE Multimedia* **2016,** *23*(2), 64–71.
19. Chowdhary, C. L.; Goyal, A.; Vasnani, B. K. Experimental Assessment of Beam Search Algorithm for Improvement in Image Caption Generation. *J Appl. Sci. Eng.* **2019,** *22*(4), 691г698.
20. Khare, N.; Devan, P.; Chowdhary, C. L.; Bhattacharya, S.; Singh, G.; Singh, S.; Yoon, B. SMO-DNN: Spider Monkey Optimization and Deep Neural Network Hybrid Classifier Model for Intrusion Detection. *Electronics* **2020,** *9*(4), 692.
21. Chowdhary, C. L.; Acharjya, D. P. Segmentation and Feature Extraction in Medical Imaging: A Systematic Review. *Procedia Comput. Sci.* **2020,** *167*, 26–36.
22. Reddy, T.; RM, S. P.; Parimala, M.; Chowdhary, C. L.; Hakak, S.; Khan, W. Z. A Deep Neural Networks Based Model for Uninterrupted Marine Environment Monitoring. *Comput. Commun.* **2020a.**
23. Chowdhary, C. L. 3D Object Recognition System Based on Local Shape Descriptors and Depth Data Analysis. *Recent Pat. Comput. Sci.* **2019,** *12*(1), 18–24.
24. Chowdhary, C. L.; Acharjya, D. P. Singular Value Decomposition–Principal Component Analysis-Based Object Recognition Approach. Bio-Inspired Computing for Image and Video Processing, 2018; p 323.
25. Reddy, G. T.; Bhattacharya, S.; Ramakrishnan, S. S.; Chowdhary, C. L.; Hakak, S.; Kaluri, R.; Reddy, M. P. K. An Ensemble based Machine Learning model for Diabetic Retinopathy Classification. In 2020 International Conference on Emerging Trends in Information Technology and Engineering, (ic-ETITE), IEEE, 2020b; pp 1–6.
26. Chowdhary, C. L. Application of Object Recognition With Shape-Index Identification and 2D Scale Invariant Feature Transform for Key-Point Detection. In Feature Dimension Reduction for Content-Based Image Identification, IGI Global, 2018; pp 218–231.
27. Shynu, P. G.; Shayan, H. M.; Chowdhary, C. L. A Fuzzy based Data Perturbation Technique for Privacy Preserved Data Mining. In 2020 International Conference on Emerging Trends in Information Technology and Engineering (ic-ETITE), IEEE, 2020; pp 1–4.
28. Benson, R.; et al. A New Transformation of 3D Models Using Chaotic Encryption Based on Arnold Cat Map. International Conference on Emerging Internetworking, Data & Web Technologies, Springer, Cham, 2019.
29. Yue, W.; et al. Design of Image Cipher Using Latin Squares. *Inform. Sci.* **2014,** *264*, 317–339.
30. Arshad; Usman; Batool, S.; Amin, M. A Novel Image Encryption Scheme Based on Walsh Compressed Quantum Spinning Chaotic Lorenz System. *Int. J Theor. Phys.* **2019,** *58*(10), 3565–3588.
31. Jun-xin, C.; et al. An Efficient Image Encryption Scheme Using Gray Code Based Permutation Approach. *Opt. Lasers Eng.* **2015,** *67*, 191–204.

CHAPTER 12

Image Synthesis with Generative Adversarial Networks (GAN)

PARVATHI R.* and PATTABIRAMAN V.

School of Computer Science and Engineering, Vellore Institute of Technology, Chennai, India

*Corresponding author. E-mail: parvathi.r@vit.ac.in

ABSTRACT

The emergence of artificial Intelligence has paved the way for numerous developments in the domain of machine vision. One of the many frameworks and algorithms which have set a benchmark for the generation of data from learned parameters is Generative Adversarial Networks. In this chapter, Generative Adversarial Networks (GANs) and similar algorithms, such as Variation Auto-Encoders (VAEs), are used to generate handwritten digits from noise. Furthermore, the training data has been visualized to gain a proper understanding of the data our model is trying to learn.

12.1 INTRODUCTION

In machine learning, generative adversarial network (GAN) is found to be stimulating recent innovation. The GANs are utilized to create new data that resembles the training data exactly and hence the name generative models. GAN accomplish degree of authenticity by matching a generator, which figures out how to deliver the objective yield with a discriminator, and out how to recognize genuine information from the yield of generator.[3] The generator tries to mislead the discriminator and the same is protected by the discriminator. "Generative" depicts a class of measurable models that diverges from discriminative models.

Generally:

- New data instances can be generated by generative model.
- Discriminative models segregate between various types of information cases.

This generative model could produce new photograph of creatures that resemble genuine creatures; hence, the working of GAN and generative models are similar. All the more officially, group of data occurrences X and set of labels Y:

- The likelihood p(X,Y) or p(X) is obtained by the generative model of the GAN architecture.
- The discriminative models catch the contingent likelihood p(Y | X).

Dissimilarity between discriminative and generative models of manually written[4] digits is shown in Figure 12.1. A generative model[5] includes the allocation of the data. For instance, the models for predicting the next word in a sequence are similar to the generative models and are more simple as compared to the GANs, in light of the fact that they assign a probability to a sequence of words.[11]

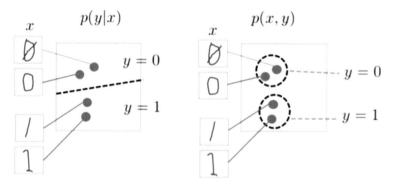

FIGURE 12.1 Generative adversarial models for handwritten digits.
Source: https://developers.google.com/machine-learning/gan/generative?hl=zh-CN. https://creativecommons.org/licenses/by/4.0/

The discriminative representation attempts to differentiate between the 0's and 1's that are handwritten, by means of a separation line drawn in the data space. On the off chance that it gets the line right, it can recognize 0's from 1's while never having to demonstrate or identify where the digits will be precisely on either side of the line. Conversely, the generative representation attempts to deliver persuading 1's and 0's by creating digits that falls near their genuine partners in the allotted data space. It needs to display dispersion all through the data space.[1,16-20]

12.2 OUTLINE OF GAN STRUCTURE

12.2.1 GENERATIVE ADVERSARIAL NETWORK

A GAN consists of two components for working:

The generator figures out how to produce conceivable data. The produced occasions become negative preparing models for the discriminator is appeared in Figure 12.2.

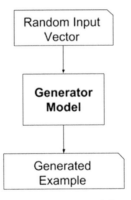

FIGURE 12.2 Example of the GAN generator model.
Source: https://developers.google.com/machine-learning/gan/generative?hl=zh-CN. https://creativecommons.org/licenses/by/4.0/

The discriminator learns to tell apart the generator's duplicate data from the original data. If the discriminator identifies the duplicate data, a penalty is imposed on the generator for producing results that can be easily identified as shown in Figure 12.3.

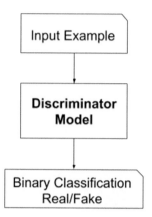

FIGURE 12.3 Example of the GAN discriminator model.
Source: https://developers.google.com/machine-learning/gan/generative?hl=zh-CN.
https://creativecommons.org/licenses/by/4.0/

When preparing starts, the generator delivers clearly counterfeit information and the discriminator rapidly figures out how to tell that it is fake.[6] As preparing advances, the generator draws nearer to creating yield that can trick the discriminator. At last, if generator preparing works out in a good way, the discriminator deteriorates at differentiating among genuine and counterfeit. It begins to group counterfeit information as genuine, and its exactness diminishes. Both the generator and the discriminator are neural systems. The input to the discriminator is directly obtained from output of the generator. Through back propagation, the discriminator's grouping gives a sign that the generator uses to refresh its loads. Generally speaking design of generative adversarial system is shown in Figure 12.4.

12.3 TRAINING DATA

12.3.1 DISCRIMINATOR TRAINING

The discriminator's preparation information originates from two sources:

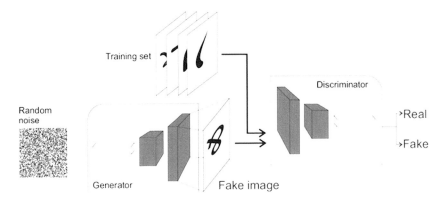

FIGURE 12.4 Architecture diagram of GAN.
Source: https://developers.google.com/machine-learning/gan/generative?hl=zh-CN

- Real information occurrences, for example, actual pictures of individuals. Discriminator utilizes the occurrences as optimistic models at the time of preparation.
- Fake information occurrences made by the generator. The discriminator utilizes the cases as negative models at the time of preparation.

Discriminator preparation steps:
1. Discriminator orders both genuine information and duplicate data from the generator.
2. The loss in the discriminator is the measure for imposing penalty for not classifying the original data as original and duplicate data as duplicate.
3. Discriminator refreshes the loads all the way through back proliferation as of the discriminator loss through the discriminator arrange.

12.3.2 THE GENERATOR TRAINING

The generator part of a GAN figures out how to make false data by joining input from the discriminator. It figures out how to cause the discriminator to group its output as real. Generator preparing requires more tight combination between the generator and the discriminator than discriminator preparing requires. The segment of the GAN that prepares the generator incorporates:

- Random input
- Generator arrange, which changes the irregular contribution to an information occasion.
- The neural network of the discriminator that is utilized for identifying the obtained data.
- Discriminator output
- Generator misfortune, which penalizes the generator for neglecting to trick the discriminator.

12.3.3 CONVERGENCE

As the generator improves with preparing, the discriminator execution deteriorates on the grounds that the discriminator will find it difficult to identify the difference between original and fake instance. In the event that the generator succeeds flawlessly, at that point, the discriminator has half exactness. In actuality, the discriminator flips a coin to make its expectation.

This movement represents an issue for combination of the GAN as a whole: the discriminator criticism gets less important after some time.[12] Even after the discriminator starts giving feedback that does not depend upon the input data, when the training of the GAN network happens continuously it ends up with generator working on the unusable feedback. This causes the quality of the generator going down and then it may collapse completely. The convergence in the GAN is often changing frequently instead of being in a stable state.[14]

12.3.4 LOSS FUNCTIONS

A GAN can have two misfortune capacities: first for generator preparing and furthermore the second for discriminator preparing. In both of those plans, in any case, the generator can just influence one term inside the separation measure: the term that mirrors the circulation of the copy information. So during generator preparing, we drop the contrary term, which mirrors the distribution of the significant data.

Minimax Loss[8] is shown in eq 11.1. The generator tries to attenuate the subsequent function while the discriminator tries to maximize it:

$$E_x\left[\log(D(x))\right] + E_z[\log(1 - D(G(z)))] \tag{12.1}$$

In the above function,

- D(x) is that the discriminator's approximation of the probability that original data instance x is real.
- Ex is the arithmetic mean over all original data instances.
- G(z) is that the generator's output when given noise z.
- D(G(z)) is the probability that a duplicate data is estimated as a original data.
- Ez is the generator's arithmetic mean of all the inputs (the arithmetic mean over all generated duplicate instances G(z)).

One of the very common issues with GANs is their high instability during training. This is because the two CNNs ideally do not take equal time for getting trained individually. This means that if the model starts to train in the wrong direction, then the latent variables do not train on the right path in the future. This leads to wrong generation and discrimination of digits. To combat this, the concept of Variation Auto-Encoders (VAEs) were introduced is shown in Figure 12.5.

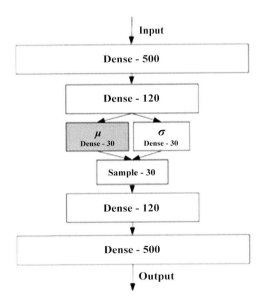

FIGURE 12.5 Architecture of a Variational Auto-Encoder.
Source: https://developers.google.com/machine-learning/gan/generative?hl=zh-CN

12.3.5 WORKING OF GANS

- The generator, that is, reversed Convolution Neural Network, accepts noise and returns a 28 × 28 image.
- The discriminator is employed with 28 × 28 images, that is, a Convolution Neural Network alongside a batch of images fetched from the real dataset, that is, the data available to us.
- The role of the discriminator is to return 0, if the generated image doesn't pass the realism check, and 1 if it does.

12.4 VARIATIONAL AUTO-ENCODERS

Similar to that of GANs, VAEs[2] are used for the representation of latent variables. The issue with auto encoders is that their latent space might not be continuous. Furthermore, there might be problems with interpolation as well. On the other hand, VAEs have latent spaces which are continuous, allowing for easy random sampling and interpolation.[10]

Due to the issue of unstability which GANs face, the model at its ideal training path, takes much lengthy time for the perfect generation of handwritten digits samples that fool the discriminator. To combat this, a ConvNet was pretrained on the MNIST dataset[7] and is utilized as a reinstatement for the previously prevailing discriminator for reducing unstability among latent variables.

One the other hand VAEs showed exception training capacity and stability. Even though it took long training periods, it reduced the chances of unstability by following the right training path and ending up with near perfect generated results. A similar concept could be applied to 3D objects stored in the format of Point Clouds. A 3D cloud can be created by compressing the data into a voxel-based compression and then fed to 3D convolutional layers instead of 2D layers.

12.5 CASE STUDY

12.5.1 DATASET

The MNIST database is a collection of 28 × 28 grayscale images of handwritten digits having a count of 60,000 training and 10,000 annotated test samples. This dataset is the smaller portion in the larger dataset provided by NIST as given by Yann LeCun, Corinna Cortes, and Christopher J.C. Burges.[4]

12.5.2 METHODOLOGY

The MNIST handwritten digits dataset contains 60,000 training and 10,000 testing images of dimension 28 × 28. There are 10 classes of data each comprising of 7000 images of the respective handwritten digits (0–9). The dataset was loaded and trained on our constructed model for over 80,000 iterations. For every iteration, the outputs were recorded and plotted to check for progress.

Trained model on the dataset for over 80,000 iterations and posted the results as shown in Figure 12.6. On applying t-SNE[9] on our dataset, as demonstrated in Figure 12.6, that at around 2000 epochs we reach a good enough clustering to understand that a generative and adversarial model would be able to pick up on the high level and low-level features as possessed by the data.

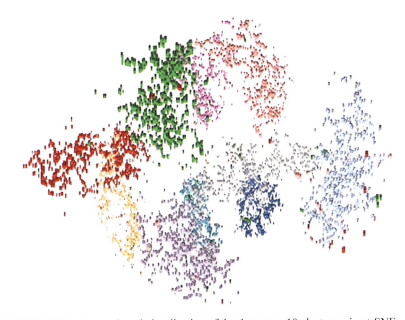

FIGURE 12.6 Tensor board visualization of the dataset as 10 clusters using t-SNE.

FIGURE 12.7 Training and generation results.

After a certain point in training time, the outliers and noise in the data gets reduced and the network starts generating near perfect visualizations of the handwritten digits in the dataset. Figure 12.6 shows how well the data were analyzed by the network and images generated.

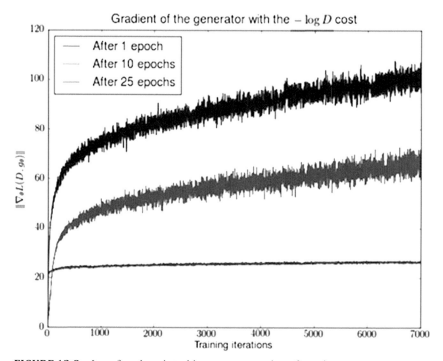

FIGURE 12.8 Loss function plot with respect to number of epochs.

Due to the issue of unstability which GANs face, the model at its ideal training path,[13] took a longer time for perfecting the model and generating handwritten digit samples that fool the discriminator. To combat this, a ConvNet was pretrained on the MNIST dataset and then used as a replacement to the already existing discriminator for reducing unstability among latent variables. One the other hand, VAEs showed exception training capacity and stability. Even though it took long training periods, it reduced the chances of unstability by following the right training path and ending up with near perfect generated results. Even though computationally intensive and sometimes unstable, generative networks hold the potential to solving many challenges in artificial intelligence.[15]

12.6 CONCLUSION AND FUTURE WORK

A similar concept could be applied to 3D objects stored in the format of Point Clouds. Compression is performed on the data to provide a 3D cloud with a voxel-based compression and then fed to 3D convolutional layers instead of 2D layers. Such a concept is applicable to many areas of research such as machine vision, speech, making biological and chemical discoveries. Even though computationally intensive and sometimes unstable, generative networks hold the potential to solving many challenges in artificial intelligence.

KEYWORDS

- **generative adversarial networks**
- **variational auto-encoder**
- **convolution neural network**
- **point clouds**
- **neural systems**

REFERENCES

1. Goodfellow, I.; Pouget-Abadie, J.; Mirza, M.; Warde-Farley, D.; Ozair, S.; Bengio, Y. Generative Adversarial Nets. In *Advances in Neural Information Processing Systems*; 2014; pp 2672–2680.
2. Kingma, D. P.; Welling, M. Stochastic Gradient VB and the Variational Auto-Encoder. In *Second International Conference on Learning Representations, ICLR*, 2014; Vol. 19.

3. Tu, Z. Learning Generative Models via Discriminative Approaches. In *2007 IEEE Conference on Computer Vision and Pattern Recognition*; 2007; pp 1–8.
4. LeCun, Y.; Cortes, C.; Burges, C. J. MNIST Handwritten Digit Database; 2010.
5. Revow, M.; Williams, C. K.; Hinton, G. E. Using Generative Models for Handwritten Digit Recognition. *IEEE Transac. Pattern Anal. Mach. Intell.* **1996**, *18*(6), 592–606.
6. Bengio, Y. Learning Deep Architectures for AI. *Found. Trends® Mach. Learn.* **2009**, *2*(1), 1–127.
7. Agarap, A. F. An Architecture Combining Convolutional Neural Network (CNN) and Support Vector Machine (SVM) for Image Classification. *arXiv preprint arXiv: 1712.03541*, 2017.
8. Berger, J. O. Admissible Minimax Estimation of a Multivariate Normal Mean with Arbitrary Quadratic Loss. *Ann. Stat.* **1976**, *4*(1), 223–226.
9. Maaten, L. V. D.; Hinton, G. Visualizing Data Using t-SNE. *J Mach. Learn. Res.* **2008**, *9*(Nov), 2579–2605.
10. Bahuleyan, H.; Mou, L.; Vechtomova, O.; Poupart, P. Variational Attention for Sequence-to-Sequence Models. *arXiv preprint arXiv:1712.08207*, 2017.
11. Sutskever, I.; Martens, J. G. E. Generating Text with Recurrent Neural Networks. In *Proceedings of the 28th international conference on machine learning (ICML-11)*, 2011; pp 1017–1024.
12. Barnett, S. A. Convergence Problems with Generative Adversarial Networks (GANS). *arXiv preprint arXiv:1806.11382*, 2018.
13. Kodali, N.; Abernethy, J.; Kira, Z. On Convergence and Stability of GANS. *arXiv preprint arXiv:1705.07215*, 2017.
14. Liu, S.; Bousquet, O.; Chaudhuri, K. Approximation and Convergence Properties of Generative Adversarial Learning. In *Advances in Neural Information Processing Systems*; 2017, pp 5545–5553.
15. Ledig, L.; Huszár, F.; Caballero, J.; Cunningham, A.; Acosta, A.; Shi, W. Photo-Realistic Single Image Super-Resolution Using a Generative Adversarial Network. In *Proceedings of the IEEE Conference on Computer Vision and Pattern Recognition*, 2017; pp 4681–4690.
16. Chowdhary, C. L.; Goyal, A.; Vasnani, B. K. Experimental Assessment of Beam Search Algorithm for Improvement in Image Caption Generation. *J. Appl. Sci. Eng.* **2019**, *22*(4), 691–698.
17. Khare, N.; Devan, P.; Chowdhary, C. L.; Bhattacharya, S.; Singh, G.; Singh, S.; Yoon, B. SMO-DNN: Spider Monkey Optimization and Deep Neural Network Hybrid Classifier Model for Intrusion Detection. *Electronics*, **2020**, *9*(4), 692.
18. Chowdhary, C. L.; Acharjya, D. P. (2020). Segmentation and Feature Extraction in Medical Imaging: A Systematic Review. *Proc. Comput. Sci.*, **2020**, *167*, 26–36.
19. Reddy, T.; Swarna Priya, R. M., S. P.; Parimala, M.; Chowdhary, C. L.; Hakak, S.; Khan, W. Z. A Deep Neural Networks Based Model for Uninterrupted Marine Environment Monitoring. Comput. Commun. **2020**, *157*, 64–75
20. Chowdhary, C. L. 3D Object Recognition System Based on Local Shape Descriptors and Depth Data Analysis. *Recent Patents Comput. Sci.* **2019**, *12*(1), 18–24.

Index

A

Aarnold cat map (AC), 228
Adaptive histogram equalization (AHE), 51
Adaptive swallow swarm optimization (ASSO) algorithm, 124
Apache Spark, 80–81
Application module
    usability test, 218
Applied adaptive wind-driven optimization (AWDO)-based multilevel thresholds, 130
Artificial intelligence as a service (AIaaS), 197
Auto insurance companies
    deep learning technique, 198
    traditional clamming process, 198
Automatic clustering (ACDE) algorithm, 127
Automatic driving, 183
AVR ratio, 45

B

Bacterial foraging optimization (BFO), 124
Bag-of-Visual-Words models (BoVW), 94
    deep learning model, 111–112
    local feature extraction, 96–98
        state-of-art local feature, 96
        vocabulary construction, 96
Bag-of-Words (BoW) model, 94
Basic intensity image (BII), 223
Batch normalization, 56
Big Data, 71–72
Big image data processing (BIDP), 69, 72
    background, 74
        IEEE BigMM Conference, 77
        surveillance videos, 77–78
        video assignments, 77
    categories, 72–74
    deep learning model, 111–112
    dimension based huge images, 75
    evolving, 73
    existing technologies
        Apache Spark, 80–81
        Hadoop, 79
        MDSC, 79–80
    future research, 87
    images, 74
    methods, 82–83
    size based huge images, 76
    technologies and implementation issues, 82–87
Bilingual Evaluation Understudy (BLEU), 155
Blood vessels (BV), 51
    retinal, 40, 41, 43
    segmentation methods, 52
Brain MR images, 121
    applications, 124
    processing, 137–139
Bright channel prior (BCP), 2, 5–6

C

Capsule neural network (CapsNet), 197, 202, 209. *see also* Intelligent Vehicle Accident Analysis (IVAA)
    accuracy, 202
    architecture, 202, 215
    computation stages, 203
Car damage evaluation systems
    background, 199–203
Chaotic map, 226
Chatbot application, 158–161
    messenger user response, 159, 160
Clustering process, 122
Color fundus picture
    vein extraction, 52
Comparison Algorithm for Navigating Digital Image Database (CANDID), 104

Consensus-based Image Description Evaluation (CIDEr), 156
Content-based image retrieval (CBIR), 93
Contextual-bag-of-words (CBOW), 106–107
Continuous chaos (CC), 226–227
Contrast Limited Adaptive Histogram Equalization (CLAHE), 2, 60–61, 124
Convolution layer, 54
Convolution Neural Network (CNN), 53, 111, 151, 184, 201. *see also* Vision-based lane and vehicle detection
  accuracy, 193
  architecture, 187–188
  flowchart, 186–187
  kinds, 152
  results and discussion, 190
    vehicle detection, 191
Cotton Wool Spots (CWS), 43
Credit rating agencies (CRA), 165
  future work, 179–180
  Glassdoor bans access, 167
  objective, 166–167
  positive sentiment reviews implementation, 171–172
  proposed system planning, 167–169
  results and discussion, 172–174
    accuracies calculated, 174–175
    limitation, 178–179
    polarity of reviews scale, 178, 179
    subjectivity of reviews scale, 176–178
    TextBlob visualization, 176
    Vader Sentiment Analyzer, 175–176
  system design, 169
    TextBlob algorithm, 170
    Vader Sentiment Analyzer, 169–170
Credit score. *see* Credit rating agencies (CRA)
Crow search algorithm (CSA), 126
Curve Partitioning Points (CPP), 104

D

Dark channel, 4
Dark Channel Prior (DCP), 2, 3–4
Data augmentation, 56–57
Dataset approach, 186
DCP-based Image Defogging algorithm, 4–5
Deep learning APIS, 204, 211
  sequence diagram, 205
Deep learning technique, 198
Deep neural networks (DNN), 53
Differential based adaptive filtering (DAF), 134
Diffusion-weighted imaging (DWI) sequences, 129
Digital image processing, 70
  application, 70
Digital retinal images for vessel extraction (DRIVE), 58
Discrete Cosine Transform (DCT), 97
Discrete wavelet transform (DWT), 126, 224
Docker engine, 207
Double BCP (DBCP), 6
Dropouts, 56
Dynamic ICP (DICP), 81

E

Evolutionary algorithms (EAs), 121
  background
    applications, 122–124
    studies, 124–136
  future research, 140–142
  research challenges, 136–140
Extension, 200

F

Faster-CNN, 152
Find pictures of Sunset, 94
Firefly algorithm (FA), 127
Fluidattenuated inversion recovery (FLAIR), 129
Fog image
  algorithm, 8
  degradation models, 3
    BCP, 5–6
    dark channel prior, 3–4
    DCP-based image defogging, 4–5
  experimental results, 8
    qualitative analysis, 10
    quantitative analysis, 9
    setup, 8–9

mathematical model, 3
proposed defogging algorithm, 6
    BCP with boundary constraints, 7
    BCP with pad image, 7–8
    boundary constraints, 7
Fully connected (FC) layer, 57

G

Generalized Llyod Algorithm (GLA), 99
Generative adversarial networks (GANs), 239, 240
    architecture diagram, 243
    case study
        methodology, 247–249
        MNIST database, 246–247
    discriminator model, 241
    generator model, 241
    outline of structure, 241
    training data
        convergence, 244
        discriminator, 242–243
        generator, 243–244
        loss functions, 244–245
    working, 246
Generic Edge Tokens (GET), 104
Google's Neural Machine Translation (GNMT) System, 154
Gray code (GC), 227
Green channel, 60

H

Hadoop, 79
Handwriting spiral test samples, 17
Haze free image, 3
Histogram of Oriented Edges (HOG), 97
Human eye, 39
    fundus, 40–42
        image, 41
        right eye image, 42
Hypertensive retinopathy (HR), 39, 40, 43–44
    clinical appearance, 44
    comparative analysis, 62
    comparison of ACC, 63
    conventional, methods, 50–52
    CWS, 43

database, 58
    AVRDB, 59
    DRIVE, 58
    INSPIRE AVR, 59
    structured analysis, 58
    VICAVR, 59
grading, 44–47
    Keith-Wagener- Barker HR Classification, 45
    SCHEIE Classification, 46
machine learning methods, 53–57
    batch normalization, 56
    convolution layer, 54
    data augmentation, 56–57
    dropouts, 56
    FC layer, 57
    padding, 55
    pixels matrix multiplied with kernel/filter matrix, 55
    pooling, 55
    stride, 54–55
performance metrics, 47–48
    accuracy, 49
    area under curve, 49
    positive predicted value, 49
    sensitivity and specificity, 49
proposed method, 59–60
    AVR calculation and HR detection, 61
    CLAHE, 60–61
    green channel, 60
    optic disc localization, 61
    output, 62
    vessel segmentation and classification, 61

I

IBM Watson, 199, 200
    architecture, 200
Image, 70
Image Cloud Processing (ICP), 81
Intelligence module, usability test, 218
Intelligent Vehicle Accident Analysis (IVAA) system, 197, 199
    Bulma CSS framework, 210
    evaluation, 214–217
        confusion matrix, 217

development platform comparison, 219
  intelligence module, 218
  model platform comparison, 219
  user satisfaction, 218–219
implementation
  CapsNet, 209
  data labeling tool, 208
  network, 210
  system overview and elements, 203
    data labeling tools, 203–204
    deep learning APIS, 204
    LINE official integration, 206
    software architecture, 206–207
    web monitoring application, 204–205
  VueJS framework, 210
  web monitoring, 210

K

Keith-Wagener-Barker model, 44–45
Kernel possibilistic c-means algorithm (IKPCM), 129
Keypoints/Interest points, 101–102
K-means Algorithm, 99
Kubernetes, 207

L

Lane detection, 184
  results and discussion, 191–192
Level set-based Chan and Vese algorithm, 128
LINE ChatBot, 213, 216
  interface, 212
LINE official integration, 206
Local Binary Pattern (LBP), 94
Local Ternary Pattern (LTP), 97

M

MATLAB with distributed computing server (MDCS), 79–80
Mean square error (MSE), 127, 231
Medical image security, 223
Metric for Evaluation of Translation with Explicit Ordering (METEOR), 156
Microsoft COCO Caption Evaluation module, 155, 156
  result, 157

Minimum cross entropy thresholding (MCET), 126
Modified fuzzy c-means (MFCM) algorithm, 124, 125
MongoDB, 207
Multiagent-consensus-MapReduce-based attribute reduction (MCMAR) algorithm, 128
Multi-objective bat algorithm, 129
Multi-objective evolutionary algorithm (MOEA), 125

N

Natural Language ToolKit (NLTK), 167
Neighborhood intuitionistic fuzzy c-means clustering algorithm with a genetic algorithm (NIFCMGA), 128
N-grams model, 95
  approaches, 100
  CBOW, 106–107
  challenges, 107, 111
  color, 103–104
  deep learning model, 111–112
  Keypoint-based, 101–102
  local feature extraction, 96–98 (see also Bag-of-Visual-Words models (BoVW))
  local patch-based, 102–103
  shape, 104–105
  visual character/pixel, 105
  visual sentence approach, 106
NLPMetric, 155
  results, 156
Novel image defogged model, 2
Number of pixels change rate (NPCR), 231–232

O

Object detection methods, 70
  background, 199–203
OpenStack, 207
Optimum boundary point detection (OBPD), 125
Order Preserving Arctangent Bin (OPABS) algorithm, 104

Index 255

P

Padding, 55
Pad-size (array A), 7
Pairwise Nearest Neighbor Algorithm (PNNA), 99
Parkinson's disease (PD), 13–15
  background, 15
  future research directions, 32–33
  handwriting tests, 16–18
    feature selection algorithms, 18
    spiral test samples, 17
  literature review, 19–26, 27–29
  solutions and recommendations, 26, 30–32
  treatment, 18
  voice data, 15–16
Pathological brain detection system (PBDS), 130
Peak signal to noise ratio (PSNR), 128, 231
Permutation techniques, 226
  arnold cat map (AC), 228
  chaotic map, 226
  continuous chaos (CC), 226–227
  gray code (GC), 227
  sudoku code (SC), 227
Pooling, 55
Proposed defogging algorithm, 6
  BCP with boundary constraints, 7
  BCP with pad image, 7–8
  boundary constraints, 7
  experimental results, 8
    qualitative analysis, 10
    quantitative analysis, 9
    setup, 8–9
Proposed partial image encryption (PIE)
  method, 228–230
  architecture, 229
  experimental results, 233–236
  performance metric, 230
    MSE, 231
    NPCR, 231–232
    SSIM, 232–233
    UACI, 231–232
    UIQ index, 232

R

Recall-Oriented Understudy for Gisting Evaluation (ROUGE), 156
Rectified Linear Unit (RELU), 187
  activation function, 190
  use of function, 187–188
Recurrent neural network (RNN), 151
Region of background (ROB), 225
Region of interest (ROI), 225
Region proposal network (RPN), 202
Region-based convolutional neural network (R-CNN), 201
Resilient Distributed Dataset (RDD), 80
REST architectural style, 209

S

Scale Invariant Feature Transform (SIFT), 97
Scene graph, 149
  background, 151–153
    CNN and RNN, 151
  chatbot application, 158–161
  evaluation, 155–158
  example, 151
  generator, 150
  methodology, 153–155
    preprocess data procedure, 154
    translation step, 155
    previous generation approach, 153
Semantic gap, 94
Semantic Propositional Image Caption Evaluation (SPICE), 156
Shift-invariant shearlet transformation (SIST), 122
Signal to noise ratio (SNR), 128
Single image defogging technique
  visual quality improvement, 1
Single shot detector (SSD), 152
Six-dimensional hyper chaos, 224
Speeded Up Robust Features (SURF), 97
Static ICP (SICP), 81
Stochastic resonance (SR), 129
Stride, 54–55
Structural similarity index measure (SSIM), 232–233

Structural similarity index metric (SSIM), 128
Sudoku code (SC), 227
Support Vector Machine (SVM), 201

T

Tangent Function (TF), 104
Template matching method, 200
TextBlob algorithm, 170
Textblob visualization, 176
Thai General Insurance Association (TGIA), 209
Tolerance rough set (TRS), 127
Tolerance roughset firefly-based quick reduct (TRSFFQR), 127
Toyota Camry data set
   accuracy, 216
Translation edit rate (TER), 156

U

Unified average changing intensity (UACI), 231–232
Universal image quality (UIQ) index, 232

V

Vader Sentiment Analyzer, 169–170, 175–176
Variation Auto-Encoders (VAEs), 239
Vector quantization, 95
Vision-based lane and vehicle detection, 183. *see also* Convolutional neural networks (CNN)
   literature study, 185
   proposed approach
      architecture, 187–188
      convolution operation, 188–190
      dataset, 186
      high level system, 188
   results and discussion, 190–194
      lane detection, 191–192
      performance measure, 193–194
Visual quality improvement
   single image defogging technique, 1
   weather conditions and the corresponding particle size, 2
Vocabulary construction, 96, 99
   GLA, 99
   K-means Algorithm, 99
   PNNA, 99
Voice data
   features, 16
VueJS framework, 210

W

Web monitoring, 210
   application, 204–205, 212
   login page, 210
   sequence diagram, 205
Word error rate (WER), 155

Y

YOLO, 152

Z

Zernike Moments, 94